*Proceedings of the Thirtieth
Annual Biology Colloquium*

The Annual Biology Colloquium

Biological Ultrastructure

Dr. Park   Dr. Luft

Dr. Nass   Dr. Cohen-Bazire

Dr. Noll   Dr. Wildman

Biological Ultrastructure:
The Origin of Cell Organelles

Proceedings of the Thirtieth Annual
Biology Colloquium, April 25-26, 1969

Edited by PATRICIA J. HARRIS

Corvallis:
Oregon State University Press

© 1971 by the Oregon State University Press

ISBN: 0-87071-169-5   LCC: 52-19235

Printed in the United States of America

Preface

SELF-REPRODUCING ORGANELLES such as mitochondria and chloroplasts have long been the subject of controversial speculations concerning their possible origin from free-living organisms. Their similarity to bacteria, with regard to their size, shape, and staining characteristics was noted by the early light microscopists, but the recent discovery of the presence of DNA and protein synthesizing systems in these organelles makes the question of their origin once again a topic of current interest. Although the purpose of this Colloquium was to bring together leading investigators in this area of cell organelles and to reexplore the question of their origin in the light of recent work in ultrastructure and biochemistry, we had no illusions that any definite answers would be forthcoming.

Perhaps an equally important result of this gathering was the demonstration that morphology and biochemistry can no longer be considered as independent studies. We are now contemplating visually the ultrastructure of macromolecules whose physical and chemical characteristics are revealed in the form of sedimentation patterns or peaks of radioactivity. No longer is it acceptable for a biochemist to announce the chemical properties of a centrifuge fraction without some visual verification of its purity, or for an electron microscopist to claim certain analogies of structure or suggestions of function without supportive chemical evidence.

Bringing together such sophisticated tools as the electron microscope, the scintillation counter, and the ultracentrifuge has accelerated our discoveries concerning the nature of living entities, but in our enthusiasm we often run the risk of overextending our interpretations of what we see. The extent of artifacts is not always fully appreciated. For this reason we invited Dr. John Luft of the Department of Biological Structure at the University of Washington Medical School to act as our Colloquium leader. Although he claims to disqualify himself as one of the "contestants proper" and alludes to himself as merely a "friend of the court," he is eminently qualified to set the stage and define the limits within which we can operate. After earning his medical degree at the University of Washington, he continued his training in electron microscopy with Dr. Don Fawcett at the Harvard Medical

School, with Drs. Keith Porter and George Palade at the Rockefeller Institute, and in European laboratories. He is one of the pioneers in biological electron microscopy, and his innovations and developments in methods of ultrastructure research are now standard procedures in electron microscope laboratories everywhere.

Among the "contestants proper" are five distinguished scientists whose work has contributed immeasurably to our knowledge of the structure and function of chloroplasts, mitochondria, and photosynthetic bacteria.

Chloroplasts have been known for some time as the organelles capable of trapping the light energy from the sun and converting it to a form usable for organisms. The efficiency with which this operation is carried out depends on the structural arrangement of the photosynthetic pigment in the chloroplast. The research interests of Dr. Roderic B. Park have been concerned with the relationship of chloroplast fine structure to the biochemical reactions of photosynthesis. Working in the laboratory of Dr. Melvin Calvin at Berkeley, following his doctoral work at the California Institute of Technology, he has brought all the armaments of biochemistry and electron microscopy to bear on the problem. Most notable of these is the technique of freeze-etching, which has revealed intra-membrane particles whose function in photosynthesis and their relationship to his previously described "quantosomes" is still the object of investigation.

The idea of autonomous organelles implies some sort of genetic mechanism for self-duplication, and it was the pioneering work of Dr. Margit M. K. Nass that demonstrated the presence of such a system in mitochondria. Dr. Nass received her Ph.D. from Columbia University, working under Dr. T. Hayashi on muscle ATP-ase. However, she left this field and joined the ranks of electron microscopists while working in the laboratory of Dr. Björn Afzelius in Stockholm. Returning to the United States, she joined the staff of the University of Pennsylvania Medical School. Her early work on mitochondrial DNA carried out with her husband Dr. Sylvan Nass at the Wenner-Gren Institute was followed by a series of elegant studies on the physical and chemical characteristics of isolated mitochondrial DNA. While these studies have shown that mitochondria and chloroplasts are not completely autonomous, in that they require genetic information from the nucleus, recent work indicates that isolated mitochondria and chloroplasts from various sources can be ingested by mouse fibroblasts in culture and survive, perhaps even function normally for many cell generations.

We turn to the work of Dr. Germaine Cohen-Bazire for some clue to the evolutionary relationship between chloroplasts and present-day photosynthetic procaryotic organisms. Dr. Cohen-Bazire was educated in France and received her doctor's degree from the Sorbonne. After working in the laboratory of Jacques Monod at the Pasteur Institute, she came to the United States to work with Dr. Roger Stanier at Berkeley on photosynthetic bacteria, and has remained there since. The organisms which she describes carry out an anaerobic type of photosynthesis and can be divided into two subgroups, purple and green, with respect to the structure and composition of their photosynthetic apparatus. The association of photopigments with unit membrane systems, characteristic of purple bacteria, also occurs in blue-green algae and in the chloroplasts of all eucaryotic groups, and probably represents the main line of structural evolution of the photosynthetic apparatus. The green bacteria, on the other hand, are unique in possessing a photopigment system that is localized in vesicles which are neither bounded by nor composed of unit membranes and which represent structurally an evolutionary dead end.

The extent to which chloroplast DNA functions in the reproduction and growth of chloroplasts has been the special interest of Dr. Samuel Wildman, professor of botany at UCLA. Dr. Wildman completed his undergraduate work at Oregon State University before obtaining his Ph.D. in botany at the University of Michigan. His methods of pursuing the question of DNA function have been varied and ingenious, utilizing mutants with defective chloroplasts and TMV-infected tobacco leaves whose chloroplast DNA transcription is shown to be regulated by the virus. The results strongly indicate that chloroplast DNA does function during growth and replication of chloroplasts, and Dr. Wildman shares with us some of his speculations concerning the nature of this functioning.

Dr. Hans Noll is a native of Basel, Switzerland, where he obtained his Ph.D. from the University of Basel. After coming to this country and holding positions first in New York and Pittsburg, he is now at Northwestern University where he holds a joint appointment between the Department of Biological Sciences and the Department of Chemistry. True to his Swiss heritage, he enjoys the challenges of fine craftsmanship as well as the challenges of the ski slope, though the former is his main concern here. By means of high resolution sucrose gradient analysis he has been able to demonstrate with extreme clarity the differences between ribosomal species, with regard to subunit size and associated RNA. There is no doubt that bacteria, mitochondria, and chloroplasts possess ribosomes in a distinctly smaller size class than cytoplasmic ribosomes of either plants or animals. Furthermore, the sizes of the RNA contained in the ribosome subunits show a steady progression, with a few anomalies, from the smallest in the bacteria, mitochondria, and chloroplasts, through the plants, invertebrates, and lower vertebrates, to the mammalian, which are largest. And here Dr. Noll has yielded to the obvious temptation to present us with a possible scheme of ribosomal evolution.

The papers presented at this Colloquium did not prove that mitochondria and chloroplasts began their intracellular existence as endosymbionts, but there is nothing here to disprove the idea and much to support it. While some people may scoff that such speculations are futile intellectual gymnastics, the implications in terms of evolutionary theory and speciation are obviously tremendous.

Special thanks for the success of the Thirtieth Annual Biology Colloquium go to Dr. Thomas C. Allen, Jr., co-chairman, and members of the Colloquium Committee, as well as to members of Phi Kappa Phi, Sigma Xi, Phi Sigma, and Omicron Nu honorary societies. A grant from the National Science Foundation contributed substantially toward making this meeting possible and is gratefully acknowledged.

PATRICIA J. HARRIS
Co-chairman, Colloquium Committee

Contents

Ultrastructure and Function With Regard to the Origin of Cell Organelles: Trials and Tribulations

JOHN H. LUFT, M.D.
Department of Biological Structure
University of Washington Medical School, Seattle

THIS COLLOQUIUM was convened to examine a problem of current interest in biology, the origin of cell organelles. I doubt that any of the contributors or those responsible for organizing the Colloquium were under the illusion that a definite answer would emerge. There are so few certainties anywhere in biology that the odds are overwhelmingly against our success in this arena. The best that we can do is to re-examine the problem in the light of current knowledge to see which of the several alternative explanations is the best justified at the moment, and perhaps to see if some new possibilities exist. Even if we fail to solve the problem, we will finish knowing considerably more than when we began.

The first step is to examine briefly the procedure by which one establishes the origin of anything, and a few examples are worth considering. "Origin" usually implies a sequence in time, although muscles have an "origin" as well as an insertion, and the intersection of the two Cartesian coordinates also is the "origin," but these meanings are irrelevant to the topic at hand. If the event in question is occurring continuously in time and is extended in space, it may be easy to trace the origin; a survey party can explore a river back to its origin. If the event occurs repeatedly in time, one need only observe closely the expected source of the event; a seed can be observed as it sprouts, leafs out, flowers, goes to seed, and dies. If the event occurred once, and through human agency, historians may be of assistance; history documents the origin of English law in the Magna Charta in 1215. When the event occurred in antiquity, the search becomes more indirect; historians, archaeologists, and materials scientists, using scientific equipment, have combined their efforts to uncover the origins of glassmaking and steel manufacture. Now, for the first time, the answer becomes probabilistic. The most reasonable explanation is that which best accommodates the largest number of those fragments of evidence which are available at the moment. An important corollary is that the answer is tentative since it may be revised whenever supplementary evidence is produced.

We have now arrived at the stage where biological origins can be added to the examples. As biologists, you are familiar with the methods which were used by Darwin and others to establish the origin of species. Now the historians are useless, and the archaeologists have moved aside to admit the paleontologists, but scientists with analytical instruments are even more necessary than before. The latest chapter reveals biochemists using protein sequence comparisons and DNA hybridization methods to complement, and often to verify, phylogeny as derived from classical biology. Since the organelles which we will be examining are present in cells of presumably advanced organisms as well as in cells considered to be primitive, it is probable that their origin was ancient. Hence, we should expect the papers presented at this Colloquium to be diversified, combining classical biology with state-of-the-art biochemistry and electron microscopy, as well as quite indirect. All of these approaches are necessary to acquire sufficient details of structure or function by which multiple similarities and comparisons may be established between the organelles in question and various free-living forms, or to formulate a plausible sequence of incorporation of a free-living organism as an endosymbiont. The greater the congruence between various properties of the free-living form and the intracellular organelle, the greater the probability of one having originated from the other. Or stated less ceremoniously, "If it looks like a duck and walks like a duck and quacks like a duck, it's a duck."

A word of caution is justified at this point. There is ample evidence from biology that the same tricks are discovered over and over again in different species with no evidence of one being derived from another. Thus, similarities themselves are not conclusive evidence for origin. Let me illustrate from personal experience: Some years ago I was interested in the electric organ of various fishes because it seemed that the electric tissue was derived from muscle. The literature search revealed that electric tissue was known in half a dozen or so varieties of fishes regarded as being widely separated taxonomically, both teleosts and elasmobranchs, but that in each case the electric tissue arose from voluntary muscle. In some it arose from longitudinal body musculature, in others from tail muscles, from jaw muscles, from external ocular muscles, or from a muscle layer in the skin. All of the biological evidence indicated that the electric tissue appeared independently in each of these fishes, with no plausible common ancestor. The histology of

many of these electric tissues was quite different, but several were nearly identical. In each case the physiological mechanism employed for generating the electricity appeared to be the same. Protein sequence comparison and DNA hybridization may be less susceptible to such confusion.

Having disposed of some of the philosophical underpinnings of the Colloquium, the next step for me is to define some of the terms used in my title "Ultrastructure and Function in Regard to the Origin of Cell Organelles: Trials and Tribulations." But, first of all, we should define Colloquium:

(1) *Colloquium:* An informal conference or group discussion.

(2) *Cell organelles:* Identifiable subunits (little organs) of the cells which have their own distinctive morphology and perform specific cellular functions. Often they are set off from the rest of the cytoplasm by a limiting membrane and they may have more membranes associated with their internal structure. The list includes:

(a) Mitochondria
(b) Chloroplasts
(c) Golgi apparatus
(d) Endoplasmic reticulum
    ergastoplasm
(e) Lysosomes and secretion
    granules

> Intimately associated
> with membranes

(f) Centrioles
(g) Chromosomes
(h) Nucleolus
(i) Ribosomes

Figure 1 illustrates the structure of several of these organelles as they would appear in an animal cell. The expanded views portray their typical structure as seen with the electron microscope. Because this represents an animal cell, the chloroplasts are not included.

(3) *Structure:* The relationship of parts to each other, the totality of which constitutes the whole.

(4) *Ultrastructure:* Structure extending to include parts which are too small to be seen directly (or be resolved) with the light microscope. Ultrastructural information comes directly from the electron microscope, and indirectly from the polarizing microscope, X-ray diffraction, and certain physico-chemical and electronic methods.

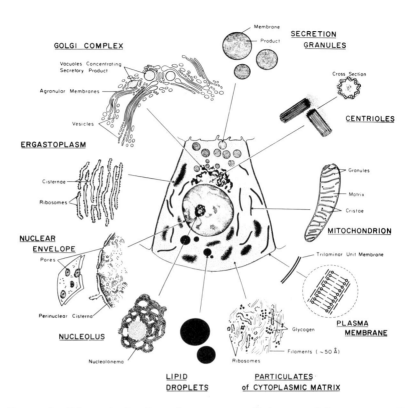

FIGURE 1.   Schematic drawings of a secretory cell (pancreas) from an animal, with exploded views of several organelles as seen in thin sections with the electron microscope. Chloroplasts are missing since this is not a plant cell. (Drawn by G. H. Turner. Reproduced by permission of W. B. Saunders Company from W. Bloom and D. W. Fawcett, *A Textbook of Histology*, 9th edition, 1968.)

(5) *Function:* The kind of action or activity proper to any person or thing; the purpose for which something is designed or exists. Note that teleology enters for the first time here.

(6) *Membrane:* A thin, pliable sheet which may or may not be porous. As applied to cells, the definition is operational, but it is consistent with the previous sentence and contains lipid, protein, and polysaccharide. The difficulties alluded to in arriving at a concise definition acceptable throughout biology epitomize the reference in the title to "trials and tribulations."

We shall begin with mitochondria, which are the best known of all the organelles in the list, to better characterize them before we inquire into their origin. Kölliker in about 1850 first recognized the large mitochondria in insect muscle, later called sarcosomes, and subsequently showed them to have a limiting membrane by the swelling which took place when he teased them into water (Cleland and Slater, 1953). In 1890, a new and relatively specific stain enabled Altmann to clearly identify mitochondria in a variety of fixed tissues. He called them "bioblasts" and because certain bacteria stained similarly, he suggested that they were free-living elementary organisms, thus anticipating the idea that mitochondria were endosymbionts, but without enough evidence to prove his point. Figure 2 shows the densely stained mitochondria in the tubular cells of mouse kidney standing out sharply from the rest of the cell contents. The tissue has been stained by Altmann's aniline-acid fuchsin mixture. The next event of significance was the discovery by Michaelis in 1898 that the dye Janus-green stained

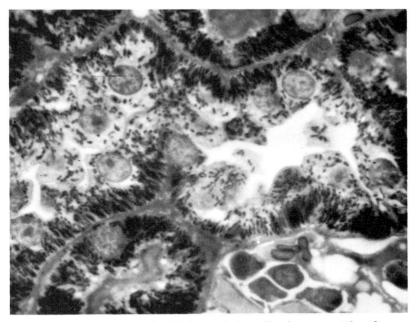

FIGURE 2.   Light micrograph of kidney tubular cells of a mouse. The Altmann stain (aniline-acid fuchsin) has stained the mitochondria a brilliant red, which appear here as dense rods. The other tissue components stain blue with the Mallory stain. Paraffin section, 2 μ thick. $\times$ 680.

living mitochondria very selectively, sufficiently so that it served to define them. Strangeways and Canti in 1927 showed by continuous observations that mitochondria were preserved by certain fixatives, particularly osmium tetroxide, in the same location and shape as they were in the living cell moments before. Their behavior, motion, fusion, and separation are now easily followed in living cells at full optical resolution by phase-contrast microscopy.

There is a parallel functional pathway for mitochondria. Nearly a century before Kölliker, Lavoisier recognized animal respiration, and knowledge about the chemical properties of tissues progressed through Claude Bernard, with Warburg in 1913 showing that cellular oxygen consumption resided in particulate elements of cells. In 1940 Albert Claude at the Rockefeller Institute began cell fractionation and by 1948 Hogeboom, Schneider, and Palade isolated liver mitochondria by differential centrifugation from an homogenate in sucrose solution. Biochemists can show that the Krebs citric acid cycle and fatty acid oxidation are located in this fraction. These reactions now serve to define mitochondria biochemically.

This history, which is nicely reviewed by Lehninger (1964), illustrates clearly the pattern by which the relationship between structure and function can be approached. The enormous amount of effort encompassed in these two paragraphs was complemented by the efforts of electron microscopists, particularly Palade, working with biochemists in showing that the internal structure of the isolated mitochondria agreed with that previously identified in mitochondria of sectioned cells. The electron microscope reciprocally was of great value in optimizing the cell fractionation procedure for purity and homogeneity of the particles which were isolated. The work is not done yet; oxidative phosphorylation, which traps the energy produced from glucose and fat oxidation as ATP, seems to be the main function of mitochondria. However, the details of the mechanism by which this is accomplished and how it relates to the lipid membranes of the mitochondrion are still virtually unknown.

Now we can approach the problem of attempting to establish the origin of mitochondria. First, what do they look like in the electron microscope? Figure 3 shows a field of mitochondria from the same tissue (mouse kidney) as Figure 2, but now with the electron microscope. Figure 4 shows these at higher magnification. The membranous internal structures of the mitochondria are now easily visible. What are the possibilities of their origin?

(1) That they arise *de novo* from components of the cytoplasm and under control of the cell's nucleus in the same way, for example, that muscle filaments arise in striated muscle cells.

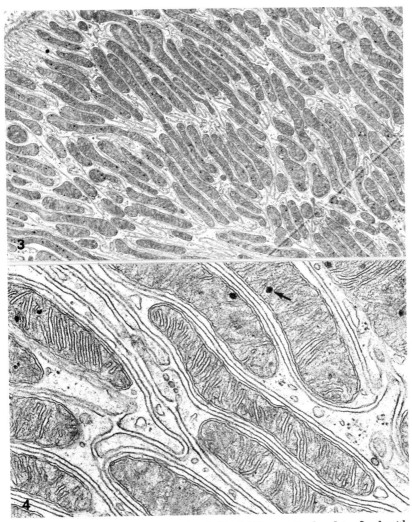

FIGURE 3. Electron micrograph of a field of kidney mitochondria, fixed with osmium tetroxide and stained with heavy metals. These mitochondria typically appear long and worm-like. × 7,000.

FIGURE 4. Same as Figure 3 but at higher magnification. Black lines representing membranes are visible everywhere. The mitochondria are bounded by an outer membrane and their internal structure (cristae) derives from an inner membrane as leaf-like folds. Mitochondria here appear to be separated from each other by meandering pairs of black lines, which in reality are deep folds of plasma membrane which greatly increase the surface of this cell (distal convoluted tubule). The small, round, very dense granules (arrow) in the mitochondria are thought to represent calcium or magnesium phosphates. × 35,000.

(2) That they arise from a pre-existing nonmitochondrial structure.

(3) That they arose originally from progressive adaptation of a symbiotic organism to the extent that they and the host cell are now mutually dependent upon each other, with the mitochondria continuing as a self-duplicating element arising from division of pre-existing mitochondria.

An example of (1) is a paper by Bell and Mühlethaler (1964) claiming the complete degeneration of mitochondria in the egg cells of a fern (*Pteridium aquilinum*) with the subsequent reappearance of a large population of healthy looking mitochondria. The mitochondria are defined here by their electron microscopic structure, which may be inadequate under unusual circumstances. There is no doubt that these mitochondria have changed dramatically, but their complete disappearance cannot be proven morphologically.

Possibility (2) arises from several observations using the electron microscope. J. F. Hartman (1954), described mitochondria in motor nerve cells as arising by accretion of submicroscopic cytoplasmic particles at or near the nucleo-cytoplasmic interface. The granules of the thickened nuclear membrane were observed to be the same size and electron density as the constituent granules of the mitochondria. Here the electron microscopy is primitive, and the size and density of particles are insufficient evidence to establish their identity.

Another paper relating to possibility (2) is that of Rouiller and Bernhard (1956) in which they followed by electron microscopy the regeneration of liver cells following CCl_4 poisoning. They suggest that dense granules in the liver termed "microbodies" are the precursors of mitochondria. DeDuve and colleagues (Leighton et al., 1968) have shown recently that liver microsomes contain oxidase enzymes quite different from those of mitochondria, and the failure of others to substantiate this proposition probably rests in their using an imprecise definition of mitochondria, or accepting minimal structural similarities as evidence of continuity.

A third example of possibility (2) is the suggestion of J. D. Robertson (1964) that mitochondria might form by a finger-like projection of cytoplasm covered by plasma membrane curving back on itself to re-enter the cytoplasm, budding off cristae, and pinching off from its stalk to form an isolated mitochondrion. In a series of drawings, Figure 5 shows the details of this concept. Aside from the fact that this process has never been substantiated by any pictures, there is a difference in the thickness of the membranes of the mitochondrion and the plasma membrane (T. Yamamoto, 1963). This suggestion is primarily a topological adventure with little relevance to mitochondrial biology.

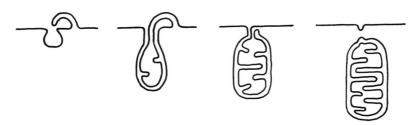

FIGURE 5.   This set of four drawings illustrates the concept of mitochondria arising from a finger-like projection of a cell re-entering itself without perforating any membranes and eventually breaking free from its membranous connection. Although topologically possible, and correct with regard to the placement and number of membranes, there is no experimental evidence to support this scheme. (Redrawn from J. D. Robertson, 1964.)

The last possibility (3), which is an elaboration of Altmann's fortuitous guess that mitochondria were endosymbionts, is now receiving substantial indirect support, and this will be amply documented by subsequent papers. If the mitochondria are self-perpetuating structures flourishing in a tolerant cytoplasmic environment, then by all that is known from molecular biology, they should have their own complement of DNA and the protein synthetic machinery in the form of ribosomes to read and follow the RNA script. Surprisingly enough, this appears to be the case, and stranger still, the DNA seems to be more like that of bacteria than that of the host cell.

Assuming that possibility (3) appears to be the most likely, what sort of evidence would be acceptable to support it? First, one might attempt to separate the guest from the host and try to culture each separately. For instance, if mitochondria could be isolated and shown to increase in number as well as incorporating tracers into protein or nucleic acid, this would be compelling evidence of their capacity for independent existence. Using mitochondria isolated from the common mushroom, Vogel and Kemper (1965) report success of such an attempt. It would be even better if the host cell could be maintained in growth as well, but the conventional isolation of mitochondria results in destruction of the host cell. In the case of the lichens, however, the ability to separate the alga from the fungus, to grow each in pure culture, and then to "synthesize" the lichen again by recombining the two to regenerate the original characteristic form of the lichen is a satisfying demonstration of symbiosis (Ahmadjian, 1966).

If the symbiosis is so intimate, or if the mutual adaptation of host to guest has progressed so far that experimental separation is impossible, a weaker form of evidence may be acceptable, such as the following: An illuminating paper, using electron microscopy to document the existence of an intracellular blue-green alga, which apparently is an endosymbiont in the cytoplasm of a flagellate, is that of Hall and Claus (1963). They document the structure of the alga sufficiently so that they feel justified in not only giving the alga a separate name, but creating for it a new family as well.

Mitochondria have different appearances in the electron microscope, depending on the type of organism in which they are found. It is possible that the extent of symbiosis between the mitochondrial guests and the host cell in mammalian cells is different as compared with more primitive cells. The following electron micrographs suggest that the biochemical accommodations provided by the mammalian kidney cells permit their guest mitochondria to "travel light," whereas mitochondria in the slime mold *Physarum polycephalum* appear to carry more "baggage." Figure 6 shows mitochondria from mouse kidney showing the typical configuration with cristae in a uniform and undifferentiated matrix without noticeable densities. Figures 7, 8, and 9 show a portion of the plasmodium of the slime mold prepared in a similar manner for the electron microscope. While the mitochondria have the usual cristae, they often show a dense structure within the matrix which is not surrounded by a membrane, as well as some small rather uniform spherical densities. These structures are nowhere to be found in mammalian cells, but they are abundant in the slime mold mitochondria. There is no proof, of course, but their appearance is suggestive of, and compatible with, their being a nucleoid with ribosomes similar to those found in bacteria.

I would like to end this presentation on a note of caution by relating some of the generally avoided problems of the field I know best—biological electron microscopy—and an example or two of personal attempts to relate ultrastructure to function in the form of physiology.

In order to finally arrive at the electron micrograph images, some of which have been and will be illustrated, it is necessary to take the living material through rather involved, lengthy processes. It is important for someone along the line to understand as well as possible what is going on during this sequence, ideally, the experimenter himself; if not, by default then, the scientific public for whom the work is published. Electron microscopy itself is not difficult—merely tedious and requiring practice; the difficult part is the interpretation of the image, which requires familiarity not only with the biology but with the methodology employed to arrive at the experimental result. It cannot be overemphasized that the analytical instrument interacts with the object being analyzed, and that an accurate interpretation of the result requires that both the object being measured and the measuring instrument be considered as an inseparable unit. The principle is no less true for biology now than it was for physics 50 years ago. When the object under observation is large, the experiment is often gentle enough so that disturbances can be ignored or are obvious. As the object in question becomes smaller, more energetic and indirect methods are required so that mistakes are easily made and may not be apparent. The extrapolation is valid and still more important down into the realm of electron microscopy as well as to molecular biology, X-ray diffraction, magnetic resonance, and optical dispersion methods where the experimental result may reveal the object only as a small ripple on the background output signal of the measuring device.

There are two principal preparation methods for electron microscopy of biological materials. The first is simpler and more direct in one sense. It is adaptable to examining large molecules which have been extracted or isolated from some cell by biochemical techniques

(Legends for figures on pages 12 and 13)

FIGURE 6.   Electron micrograph of several mitochondria of kidney cells fixed with glutaraldehyde followed by osmium tetroxide, but otherwise handled the same as Figures 3 and 4. The intramitochondrial membranes (cristae) are thinner and somewhat angulated. × 35,000.

FIGURE 7.   Electron micrograph of a piece of slime mold, *Physarum polycephalum,* fixed as for Figure 6. Two large nuclei (N) dominate the picture, but in the cytoplasm are seen lipid droplets (L) and mitochondria (M). These mitochondria are round, vary considerably in density from one to another, and usually contain an irregular body of moderate density, quite different from those in Figure 4. The rectangle shows the area enlarged for Figure 8. × 7,000.

FIGURE 8.   Area within the rectangle of Figure 7. The cristae of the mitochondria now appear to be tubular, like spaghetti, instead of leaves or plates as in Figures 4 and 6. In addition, between the tubules are very small dense granules, similar in size and appearance to the ribosomes which can be seen in the cytoplasm between mitochondria (arrow). The irregular, moderately dense bodies have a very fine granular texture. The rectangle shows the area enlarged for Figure 9. × 35,000.

FIGURE 9.   Further enlargement of the area within rectangle of Figure 8. Structures of the size and staining characteristics of ribosomes are visible inside the mitochondria as well as outside in the cytoplasm (arrows). The irregular, moderately dense body appears to be finely granular, without other distinguishing characteristics. × 160,000.

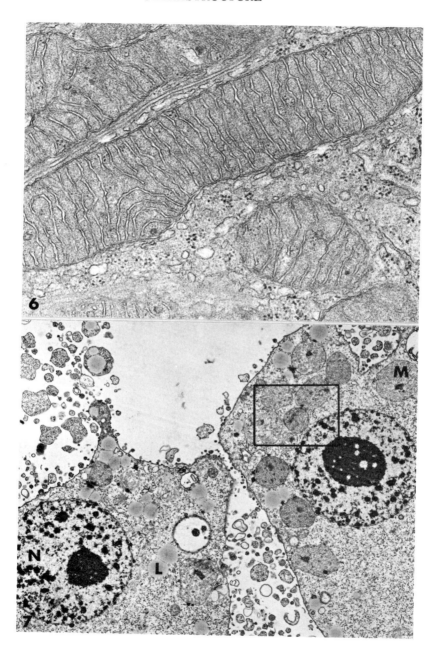

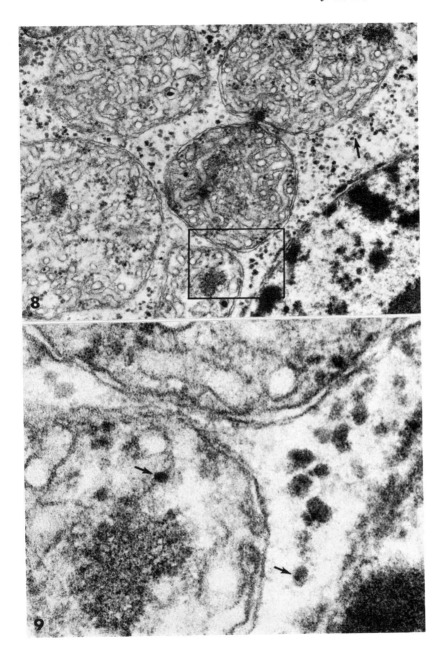

and which are available in rather concentrated and pure form. Micrographs of circular DNA result from this procedure. It involves first spreading the purified material as a thin film by allowing a drop of the solution to expand on the surface of water and touching the specimen support screen to this surface. Another variation allows a drop of the very dilute suspension of the molecules to dry on the surface of the screen. Next comes some procedure to increase the contrast of the object so that it will be visible in the electron microscope, usually by evaporating under vacuum a thin layer of heavy metal at an angle over the object, or by surrounding but not covering it with a structureless dense material (negative staining). These procedures are relatively direct and simple, but they can be no more gentle than the biochemical procedures used for the original isolation, and, of course, the structural relationship of the molecules to other parts of the cell is obliterated. And the actual observation of the preparation in the electron microscope destroys it by quickly reducing it to a delicate graphitic ash, but its outline persists as a metallized crust, or as a microfossil in the residue of the negative stain.

The second method is used to obtain sections of tissues which retain the histological relationship of cells and their contents to each other as well as preserving ultrastructure. The first and most important step is *fixation* in order to stabilize the structure throughout the subsequent operations. Unfortunately, this is the step about which the least is known. Ideally, it should be faster than the usual biological events—a millisecond or so—but fixation for electron microscopy usually takes seconds at least and perhaps minutes. The events which transpire during this time are obscure to say the least. The mechanism of fixation is probably a cross-linking of various molecules, but beyond this generality the only fact is that the successful fixatives are among the most poisonous chemicals in existence. However, the converse is not true; many highly toxic compounds are not fixatives. Among the possible physical agents which might be considered as fixatives, only freezing has been used. Although this can be faster than chemical fixation, there are difficulties which severely limit its usefulness (Rebhun, 1965).

When successfully fixed, the tissue is dead but "stabilized," and the next requirement is to slice it thin enough to examine in the electron microscope, i.e., thinner than 0.1 μ. To accomplish this, the tissue must be supported by some matrix so that it will not crush during cutting, and the usual procedure is to replace the water in the tissue and impregnate it with a liquid plastic which is later hardened. Several plastics can be used, but the epoxy resins, which react with the tissue and chemically incorporate it into the resin polymer, have been particularly successful. The plastic block with the tissue, no larger than 1 mm on a side, is sliced in a precision microtome with a piece of broken glass as a knife (or a ground and polished diamond edge) about $1/20\,\mu$ or 500 Å thick onto a water surface. These slices or sections can be stretched over the holes in a piece of 200-mesh screen, which is the specimen support, and dried and examined in the electron microscope. However, very little is seen in the electron microscope unless the section is first stained.

The staining procedure is not much better understood than is the fixation procedure, but at least there is no premium on time because the slice of tissue is rigidly held in the plastic. The electron microscope sees mass only, and a gram of water appears just as dense to electrons as a gram of lead, provided that they are spread over the same area. Because there is no selectivity for particular elements in the electron microscope, in anything approaching the way that dyes absorb light so strongly as compared to colorless organic compounds, the electron stains only need to deposit as much mass as possible in or on the tissue components. The most successful stains have been solutions of lead or uranyl salts that have their pH adjusted to be just on the verge of precipitation. There is a weak selectivity of these stains for certain tissue elements, but nothing approaching the differential staining possible for light microscopy. When all goes well, one emerges with an electron micrograph of the tissue.

The interpretation of the micrograph, that is, how successfully one can identify structures in the micrograph which correspond to reality in the living cell, is the heart of the problem. To illustrate where the difficulties lie, we shall examine one element which recurs constantly in the organelles around which these colloquium papers are organized, i.e., membranes. If it can be maintained that "the cell is a highly abstract concept of minimal content" (Picken, 1960), the situation is still worse for membranes. A good electron micrograph shows the cell to be bounded by a black line, and at higher magnification this single line can be decomposed into two, separated by a light zone and sometimes ignobly referred to as "railroad tracks." Because of a variety of difficulties, it is impossible to say how thick the "membrane" is ; the best one can do is specify the center-to-center distance between the two lines, but this can be measured with precision to be 52.00 Å $\mp$ 2.16 Å in nerve cells after a certain preparation method (Yamamoto, 1963). Figure 10 shows the appearance of the plasma membrane from an intestinal cell of a mouse. The two parallel black lines separated by a light space are typical of the "unit membrane" or "trilaminar membrane" as commonly pictured in electron micrographs. Even this simple image is deceptive because whether one sees two black lines or not depends on the focus of the microscope.

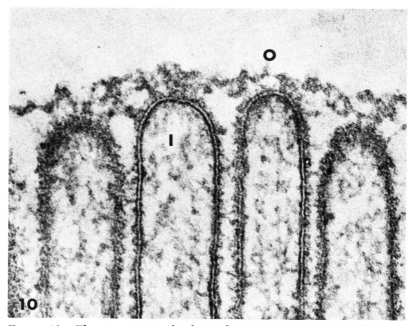

FIGURE 10.   Electron micrograph of cytoplasmic projections (microvilli) from the intestinal cell of a mouse to illustrate the twin dense lines separated by a lighter space, which is the pattern seen so frequently in cell membranes as to invoke the term "unit membrane." The cytoplasm or inside of the cell is labeled "I", and the intestinal lumen or outside is designated "O." Osmium-fixed, thin (300 Å) section is stained with uranyl acetate and alkaline lead solutions. × 160,000.

The following is a through-focus series of the plasma membrane of *Amoeba proteus*. The series of five pictures (Figures 11a-e) show the same region of the plasma membrane of the amoeba, as the lens of the electron microscope is varied in strength from too weak (under-focus) to too strong (overfocus). The "in focus" image is the center of the series (Figure 11c). Figure 11a shows the parallel black lines with a superimposed granularity which modulates the apparent structure of the "railroad tracks" as well as producing a mottled appearance on the other structures in the picture. This granularity is an image phenomenon associated with the degree of defocus and is completely unrelated to the biological question. Figure 11b is closer to focus, and the two parallel black lines separated by the single central light region are still visible, although the contrast is reduced. Figure 11c is very

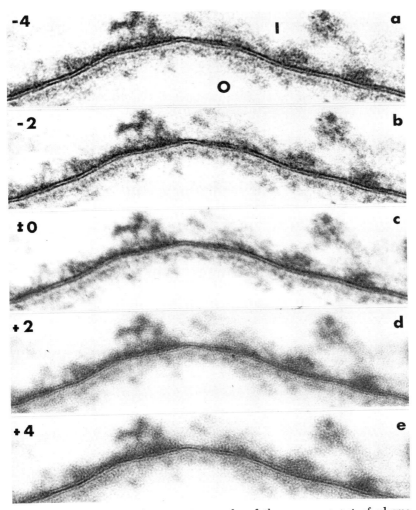

FIGURE 11. Successive electron micrographs of the same segment of plasma membrane from *Amoeba proteus,* separating the outside environment (O) from the inside cytoplasm (I). The series illustrates the changes in appearance of the unit membrane as the focus changes in the electron microscope. The figures identify approximately the number of microns by which the focal length of the objective lens differs from the true (Gaussian) focus. The contrast is high when the image is underfocused (−4 μ) but there is an artifactual granularity to the image. Closer to focus (−2 μ) the granularity is less, but the contrast is less also. These are the usual conditions under which biological electron microscopy is done, where Fresnel fringes reinforce the intrinsic contrast of the object. The

in-focus picture (c) has very low contrast, but it faithfully represents actual mass distribution in the section. In the overfocus pictures ($+2$ and $+4\mu$), the granularity returns as well as the contrast, but the fringes now antagonize the mass density of the unit membrane, and by $+4 \mu$ defocus (e) the fringes not only obliterate the trilaminar structure, but invert it to produce a five-layered "membrane" with two light layers between three dark layers, the center layer being the darkest of all. Acrolein fixation followed by osmium, uranyl, and lead staining. Preparation and micrographs kindly contributed by Dr. B. Szubinska. × 160,000.

close to focus. The "railroad track" is still visible, although contrast is at a minimum. The phase contrast granularity which accentuated the contrast in Figures 11a and 11b is likewise gone, so that the density distribution seen in this (and only this) figure corresponds quite closely to the actual mass distribution in the specimen itself. Figure 11d is as much "overfocus" as Figure 11b was "underfocus." The granularity has returned, but the image of the plasma membrane has completely changed—in fact, it has reversed itself. There are now three black lines enclosing two light zones, and the central line is black, not white. This pattern is seen more clearly in Figure 11e, with correspondingly greater granularity. The explanation is that the phase contrast phenomenon, which is responsible for the granularity, has reinforced the basic image of the "railroad track" on the underfocus side (Figures 11a and 11b) to give the pleasing high contrast seen there. On the overfocus side (Figures 11d and 11e), the phase contrast phenomena have reversed their contribution so that they now oppose the real (?) image of the "railroad tracks" with such intensity that they totally obliterate and even invert it. Unfortunately, it is usually not possible to tell in advance where the in-focus lens setting occurs without taking a through-focus series, and the in-focus picture is seldom published because it looks "fuzzy" with low contrast. Worse yet, it is possible by photographic manipulation in printing the micrograph, to deliberately suppress the granularity in the background, which is the only clue that the observer has to estimate the focus deviation at which the picture was taken. Thus, considerable variability is available to the electron microscopist which he can consciously or unconsciously manipulate in obtaining his images. These uncertainties at the electron microscope level are further compounded by the changes which are produced in membrane spacing by using various agents for the initial fixation of the tissue. Table 1 lists the variability which is found after using several conventional fixation techniques. There is no way to tell which is the "correct" value for the living cell. At present, all of these difficulties contribute to the dispute among electron microscopists concerning "the structure" of the plasmalemma.

Table 1.   Summary of Membrane Measurements*

	plasma membrane	mitochondria	ribosome-coated endoplasmic reticulum	vesicotubules	synaptic vesicles
Permanganate	51±7 (23)	41±5 (82)	44±4 (32)	51±7 (189)	—
Osmic	54±7 (104)	41±7 (22)	46±6 (12)	53±6 (113)	—
Glutaraldehyde	69±4 (31)	45±3 (97)	45±4 (25)	65±7 (232)	—
Acrolein	62±7 (132)	50±10 (51)	47±7 (26)	61±7 (97)	—
Yamamoto	52±2	44±3	46±2	—	51±2

(n) = number of measurements

* Measurements of the center-to-center distance in Angstroms between the black lines for various animal cell membranes, using a variety of fixation methods. Vesicotubules are a membrane system in the oxyntic (parietal cell) of the stomach. There is considerable variation in spacing, and the "correct" value is not known. (From C. Lillibridge, 1968.)

Most electron microscopists would call the "railroad tracks" the cell membrane, plasma membrane, or plasmalemma. Similar membranes are seen to limit most of the organelles of the cell as well as being incorporated into their structure by infolding, but the spacing is narrower. Under conditions where the plasma membrane measures 52 Å, the mitochondrial membranes, both internal and external, measure about 44 Å (Yamamoto, 1963; Lillibridge, 1968).

Electron micrographs are stationary images of structures which were arrested at some phase of their existence. If we wish to discover something of the function of this membrane, we ask physiologists, tissue culturists, or biochemists. The physiologists usually are disdainful of black lines; for them the membrane is a boundary of some sort across which ionic concentration differences exist with a corresponding electrical potential of about 0.1 volt. If they lower a microelectrode, with a tip about 1 μ in diameter, toward a muscle cell for example, they observe no voltage while the electrode is still in the Ringer's solution surrounding the cell. With a low-power microscope, it can be seen that as the electrode touches the cell, irregular low voltages are recorded on the electrode, and with further insertion, a stable potential of 80 to

90 mv suddenly appears. Most electrophysiologists would say that at that instant the microelectrode penetrated "the membrane." The experimental conditions necessary to handle the microelectrode make good microscopic observation difficult, so that it is hard to determine just where the electrode tip is located within a few microns, but the sudden appearance of the membrane potential on the electrode is an event very easily observed. Those functions of the cell membrane which are electrical (resting potential, nerve impulse, muscle action potential, hyper- and hypopolarization, etc.) can be located to perhaps $\pm 5 \mu$ of the optical surface of the cell.

If we now wish to use the full resolution of the light microscope on living cells as they touch and crawl over one another, we approach the scientists who use cell and tissue culture methods. Their cultures can be made thin enough so that the phase contrast microscope can show detail in living cells down to the resolution limit of about 0.25μ. In time-lapse movies we can follow the advancing pseudopod of a moving cell. In phase contrast the cytoplasm of the pseudopod is dark gray on the light gray background with a "sharp" boundary between the two, and the boundary advances with the cell. On further enlargement, this boundary, which ought to be "the membrane," is revealed as a gradual transition from light to dark over a 0.25μ distance. From the optical theory behind phase contrast, we can say that there is a steep gradient or discontinuity in refractive index, or better yet, in optical path length, somewhere within that 2,500 Å zone, but we cannot say how thick it is, or even if it lies in the center of the optical gradient. In an electron micrograph 2,500 Å is a large distance, but it is impossibly precise for the mechanical advance of the physiologist's microelectrode.

Last, and perhaps least, with respect to the accuracy of localization of the plasma membrane, we can consult the biochemist. Perhaps he can isolate the membrane in some fraction so that we can discover its composition and enzyme complement and perhaps deduce something about function from this information. The biochemist can, indeed, separate a number of fractions from homogenized liver cells with different properties and composition. But which of these, if any, is the plasma membrane? He replies that the fraction which has the highest activity of the enzyme 5'-nucleotidase is the plasma membrane fraction, and the higher the enzyme activity the purer the fraction because this enzyme is associated with plasma membranes (Bosmann et al., 1968). Where one is, the other should be. It is not obvious from first principles why this assertion should be true, but it is nevertheless useful to the biochemist.

Now the purpose of this detour was to illustrate the point that each specialty of biological investigation has its own definition of "plasma membrane." Furthermore, these definitions are all operational, i.e., the plasma membrane is found or detected after a certain operation is performed. Also note that the operations specified involve instruments—the instruments or methods which yield a certain kind of information around which that particular specialty was organized in the first place. In the final analysis, all of these definitions are circular, but at least they are useful within that particular field since they are reproducible. The difficulty comes in attempting to establish a dictionary which will translate the information acquired in one field into that of another; there are no simple equivalents. A corollary of this premise is expressed in the difficulty which we who teach in medical schools are now experiencing in attempting to provide the so-called "integrated courses" specified in various "core curricula" which are currently fashionable; it is not easy to integrate information from the various disciplines into one coherent package.

This basic difficulty of reliable definitions in biology was forcefully brought to my attention during a year in which I, as a well-trained histologist and electron microscopist, spent a year in a first-rate physiological laboratory in London. After some months, I learned how to dissect out good frog sartorius nerve-muscle preparations. They behaved as well as those prepared by the permanent members of the laboratory, i.e., there were few, if any, damaged muscle fibers, the muscle twitched uniformly and repeatedly, and the nerve action potential went from threshold to maximal response with a small increase in stimulus strength. One day curiosity prevailed over conformity and I fixed and embedded one of my nerve-muscle preparations and compared it with a similar preparation prepared from the nerve and muscle fixed *in situ* in the frog without exposure to Ringer's solution. When thin sections of both were examined in the light-microscope, there were striking differences. I showed the two slides to the head of the laboratory, but he objected to the muscle cross section fixed *in situ*, in particular, the polygonal, irregular profiles of the muscle cells. He much preferred the completely round muscle cell cross sections of the dissected muscle because of the uniform geometry and the large extracellular space, despite the fact that it must have resulted from swelling in the amphibian Ringer solution from the original, very compact muscle tissue with minimal extracellular space seen in control tissue. The electron microscope revealed shocking morphological damage to the myelinated nerve in the isolated preparation, with folded, shattered myelin and partial separation of axoplasm from the myelin sheath. But the nerve with its muscle was a perfectly good physiological preparation, no different from those on which 50 years of reliable physiology has been based. Were the physiologists wrong, or were the morphologists in error? The answer is neither are "wrong" and both are equally "correct." The essential point is that they both use operational standards against which to compare their experimental work, but the operations are completely different.

For the morphologist, tissue fixed as gently as possible with the least disturbance beforehand provides the standard. For the physiologist, a nerve which conducts properly and a muscle which twitches normally is the standard. A very incomplete translation between these two is perhaps possible; since the muscle twitches although swollen, and the nerve conducts with its shattered myelin, it must be that these changes are relatively unimportant for nerve and muscle function during the short time over which the physiological experiments are conducted, or that the preparations can somehow compensate for these alterations. In the long run, or for maintenance of the tissue or replacement of energy reserves, it may be that these morphological alterations would reveal a functional deficiency, but no one is certain because the experiments required to document this suggestion are difficult, if not impossible, to carry out.

During this presentation, an attempt has been made to do two things: first, to provide some background information which may be helpful in considering the problem of the origin of cell organelles, which the following papers will amplify; second, to emphasize some of the difficulties along the way at attempting to utilize or "integrate" information from the various branches of biology. It is true that we are all working on the same thing, that is, on living cells, but each of us receives our answers in the language appropriate to the instruments which we use to ask the questions. Unfortunately, there is no dictionary which will reliably translate one language into another.

ACKNOWLEDGMENTS: The author is grateful to Dr. B. Szubinska for contributing Figure 11 and to Dr. C. Lillibridge for permission to use his data for Table 1. Figure 1 is reprinted from *A Textbook of Histology* by permission of the publishers. This presentation was assisted, in part, by USPHS grants HE-16598 and NB-00401.

Literature Cited

Ahmadjian, V. 1966. Lichens. Chapter 2, pp. 35-97 in *Symbiosis*, Volume I. Associations of microorganisms, plants and marine organisms. S. Henry, ed. New York: Academic Press.

Altmann, R. 1890. *Die Elementarorganismen*. Leipzig: von Veit and Co.

Bell, P. R., and K. Mühlethaler. 1964. The degeneration and reappearance of mitochondria in the egg cells of a plant. J. Cell Biol., *20:*235-248.

Bosmann, H. B., A. Hagopian, and E. Eylar. 1968. Cellular membranes: The isolation and characterization of the plasma and smooth membranes of HeLa cells. Arch. Biochem. Biophys., *128:*51-69.

Cleland, K. W. and E. Slater. 1953. The sarcosomes of heart muscles. Their isolation, structure, and behavior under various conditions. Quart. J. Micro. Sci., *94:*329-346.

Hall, W. T., and G. Claus. 1963. Ultrastructural studies on the blue-green algal symbiont in *Cyanophora paradoxa* Korschikoff. J. Cell Biol., *19:*551-563.

Hartman, J. F. 1954. Electron microscopy of motor nerve cells following section of axons. Anat. Rec., *118:*19-33.

Lehninger, A. L. 1964. *The Mitochondrion*. New York: W. A. Benjamin.

Leighton, F., B. Poole, H. Beaufay, P. Baudhuin, J. Coffey, S. Fowler, and C. DeDuve. 1968. The large-scale separation of peroxisomes, mitochondria and lysosomes from the livers of rats injected with Triton WR-1339. J. Cell Biol., *37:*482-513.

Lillibridge, C. 1968. Electron microscope measurements of the thickness of various membranes in oxyntic cells from frog stomach. J. Ultrastruct. Res., *23:*243-259.

Picken, L. E. 1960. *The Organization of Cells and Other Organisms*, page 3. Oxford: The Clarendon Press.

Rebhun, L. 1965. Freeze substitution: Fine structure as a function of water concentration in cells. Fed. Proc., *24,* suppl. 15, S217-S232.

Robertson, J. D. 1964. Unit membranes: A review with recent new studies of experimental alterations and a new subunit structure in synaptic membranes. Pages 1-81 in *Cellular Membranes in Development*, M. Locke, ed. New York: Academic Press.

Rouiller, C., and W. Bernhard. 1956. "Microbodies" and the problem of mitochondrial regeneration in liver cells. J. Biophys. Biochem. Cytol., *2 (4)* suppl.: 355-360.

Strangeways, T. S., and R. Canti. 1927. The living cell *in vitro* as shown by darkground illumination and the changes induced in such cells by fixing reagents. Quart. J. Micro. Sci., *71:*1-14.

Vogel, F. S., and L. Kemper. 1965. Structural and functional characteristics of deoxyribonucleic acid-rich mitochondria from the common meadow mushroom, *Agaricus campestris*. II. Extracellular cultures. Lab. Invest., *14:* 1868-1893.

Yamamoto, T. 1963. On the thickness of the unit membrane. J. Cell Biol., *17:* 413-421.

The Architecture of Photosynthesis

R. B. PARK
Botany Department
University of California, Berkeley

MANY FAMILIAR PROCESSES in the biosphere are the result of sunlight absorption and its direct conversion to heat. This heat is eventually re-radiated, the total heat engine yielding wind, waves, evaporation, and precipitation. Sunlight absorption followed by direct conversion to chemical energy rather than heat also occurs on a large scale in the biosphere. However, this process is largely restricted to organisms. That certain creatures can capture and store the energy of sunlight as chemical bond energy before its ultimate degradation to heat results from structural and chemical restraints imposed on the absorption process by the organism. Organisms which possess the particular structures suited for capture of sunlight and its conversion to chemical bond energy are termed photosynthetic, and the process itself is called photosynthesis. This process occurs in procaryotic as well as eucaryotic cells, but it is eucaryotic cells which are most important in biosphere energetics. Though procaryotic cells are not a major factor in overall energy conversion, they are interesting creatures in their own right and are very useful in the study of photosynthetic mechanisms and structures. Dr. Cohen-Bazire will discuss these organisms later in this volume. This paper will be limited to a discussion of photosynthesis in eucaryotic cells.

The eucaryotic photosynthetic process is often represented as:

$$2H_2O + CO_2 \xrightarrow[\text{chlorophyll}]{\text{light}} (CH_2O) + H_2O + O_2.$$ This reaction sequence has $\triangle G$ of about $+ 120,000$ cal depending on the carbohydrate product and consists of the pumping of electrons from water to CO_2. The light-driven electron pump produces positive and negative sites within the organism. The positive site is a powerful enough oxidant to remove electrons from water with the production of oxygen and therefore has an oxidation potential of at least $+0.8$ volts at pH 7.

The negative site is a powerful enough reductant to donate electrons to ferredoxin, or to hydrogen ions to produce free hydrogen gas. Thus we find the most powerful oxidants and reductants in biology produced in the light reactions of eucaryotic photosynthesis. All other biologically produced oxidants and reductants, with the possible exception of nitrogenase in bacteria, fall between these two.

The heterotrophic organisms of the biosphere derive energy from transport of electrons from the reduced products of photosynthesis back eventually to molecular oxygen. In this sense the action of an aerobic muscle or gasoline engine are comparable. Both derive energy by electron transport from reduced products of photosynthesis to molecular oxygen.

Is the photosynthetic process localized within eucaryotic cells and if so where? The answers to this question began to appear almost 100 years ago with the classic studies of Engelmann (8).[1] This Swiss worker knew that light was the source of energy required to drive the photosynthetic process, and he designed a microscope condenser which confined the radiant energy of the light beam to a very small spot. He also possessed a strain of bacteria which moved in liquid medium toward regions of increasing oxygen concentration. By mixing these bacteria with small photosynthetic organisms on a microscope slide, he could then use the special condenser to irradiate certain parts of the photosynthetic cells and find which part must be illuminated to obtain oxygen evolution. The results of his experiments on the green alga *Spirogyra* are shown in Figure 1. Here it is seen that illumination of the chloroplast was necessary for oxygen evolution and that oxygen was evolved near the site of light absorption. Engelmann in another clever experiment showed that red and blue light were more effective in oxygen production than green, thus indicating that chlorophyll was the effective absorber of the radiant energy. But where was the site of CO_2 fixation? Was it also in the chloroplast or was it out in the plant cytoplasm? This question was not answered in a qualitative way until 1954 when Arnon, Whatley, and Allen (1) showed that small amounts of light-driven CO_2 fixation could be obtained in isolated spinach chloroplasts. The maximum fixation rates were however only a fraction of a percent of the *in vivo* rates of photosynthesis on a chlorophyll basis and could not answer whether all light-dependent CO_2 fixation was chloroplast localized. This dilemma has been solved in the past few years with improved isolation techniques which yield spinach chloroplasts with rates comparable to *in vivo* rates (11). We can now say with confidence that spinach chloroplasts are autonomous photosynthetic

---

[1] Italic numbers in parentheses refer to Literature Cited, page 39.

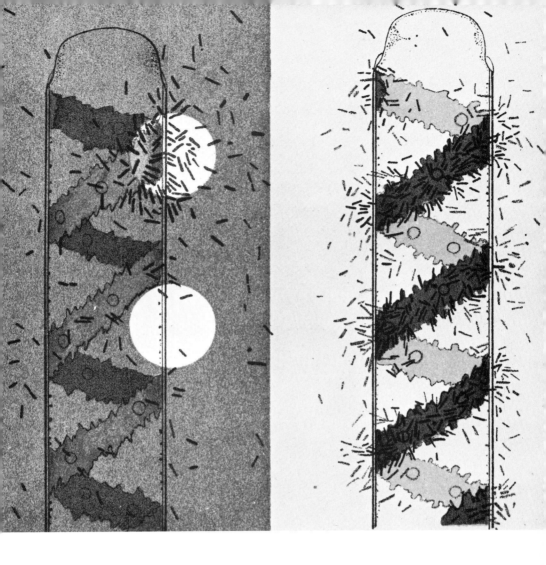

FIGURE 1. Engelmann's experiment in which he showed that the site of O_2 evolution is the chloroplast and that O_2 evolution occurs close to the region of light absorption (8).

organelles. Perhaps the most remarkable evidence for this fact comes from the observation that naked chloroplasts naturally occur in the cytoplasm of certain animal cells. Work from Leonard Muscatine's laboratory at UCLA has shown that naked chloroplasts in the digestive cells of certain marine nudibranches are not only photosynthetic, but persist for long periods of time (19).

Our problem then is to relate the process of photosynthesis with the structure of the chloroplast. We have already spoken about the process as light-driven electron transport from water to CO_2. A more detailed presentation of this pathway is given in Figure 2. This diagram summarizes not only some of our latest knowledge, but also some of our most recent errors in the studies of photosynthesis. It is generally conceded with one notable exception (2) that eucaryotic photosynthetic electron transport consists of two light reactions separated by an electron transport chain. A phosphorylation step occurs on the electron transport chain and possibly also on the cyclic path of electron transport leading from the strong photosystem 1 reductant back into the chain. The ATP and reduced NADP produced by this light-driven electron pump are used in reactions of CO_2 fixation diagrammed in Figure 2. Several years ago we would have said the path of CO_2 fixation, shown here as derived from the work of Calvin and co-workers, is the only important photosynthetic CO_2 fixation path in nature. Work by Hatch and Slack (10) during the past two years has shown that this is not the case. What we usually call sugar (cane sugar) is not made by the Calvin path at all, but apparently by a C_3 acceptor reacting with a CO_2 molecule to yield a C_4 acid. At this point we cannot say how widely distributed this new pathway is, but it shows once again how we must continue to regard apparently established fact as only a model and not the final answer. A more artistic, and with respect to the carbon cycle, more correct presentation of photosynthesis is given in Figure 3. This diagram (6) works equally well for the Calvin or for the Hatch-Slack pathway.

Our problem is to associate this machinery with known structures inside the chloroplasts. It should be remarked, of course, that a chloroplast is concerned with much more than photosynthesis, and that those parts of the chloroplast dealing with "blueprints" rather than "photosynthetic machinery" will be considered by other contributors. I shall first consider some features of chloroplast ultrastructure and then attempt to associate these with photosynthetic function.

Figure 4 is an electron micrograph of a cross section of a spinach chloroplast. It shows the general features of chloroplast structure, namely an internal membrane phase, also called thylakoids or lamellae, suspended in a colorless stroma phase. Various separation experiments have shown that the electron transport reactions leading from water to ferredoxin are localized in the green thylakoids, whereas the CO_2 fixation reactions are associated with structures in the stroma (17, 18). The chloroplast is surrounded by two membranes. Though these are often referred to as the chloroplast "double membrane," it is appealing

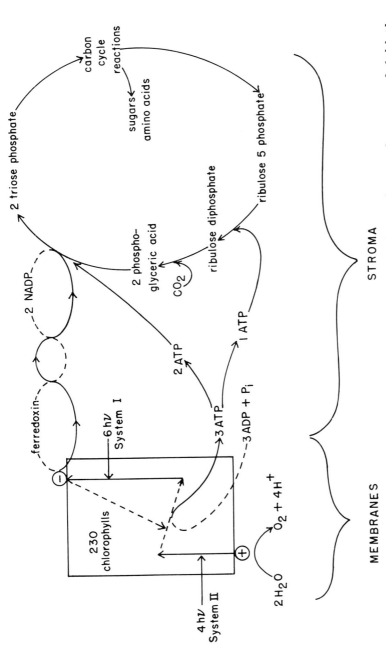

FIGURE 2. A photosynthetic scheme showing the distribution of photo-synthetic function between the stroma and thylakoid phases of the chloroplast.

Mystery of the Green Bodies

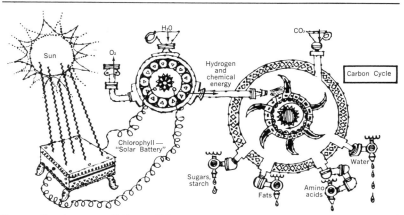

FIGURE 3.   A more artistic, and in some ways more accurate, representation of the scheme presented in Figure 2 (6). Reprinted by permission of Scholastic Magazines, Inc. from *Senior Science,* © 1967 by Scholastic Magazines, Inc.

to think that only the inner membrane belongs to the chloroplast, whereas the outer membrane is the exclusion membrane of the host cell.

In recent years the method of freeze-etching (5) has been applied to the study of chloroplasts. Here chloroplasts in aqueous suspension are frozen and fractured. Then a replica is made of the fractured surface either before or after water has been sublimed from the fractured material. A comparison of replicas from isolated spinach chloroplasts with high and low CO_2 fixation rates is given in Figures 5 and 6. The intact chloroplast (Figure 5) which yields high fixation rates is similar in appearance in many ways to the chloroplast in Figure 4. Detail is much more evident in the swollen, low fixation-rate chloroplast.

In the swollen chloroplast the stroma region is seen to contain numerous particles ranging up to 200Å in diameter. The largest of these particles are probably ribosomes and the enzyme carboxydismutase. Thylakoids seen in cross section in this micrograph are very similar in appearance to those in typical thin sections such as the ones shown in Figure 4. However, breaks along the membrane indicate a number of associated particles. That these particles are not typical stroma particles is readily indicated by centrifugally washing spinach thylakoids in dilute buffer. The washing treatment removes the soluble stroma particles and leaves fairly clean thylakoids. After the thylakoids are washed, fracturing indicates the continued presence of membrane-

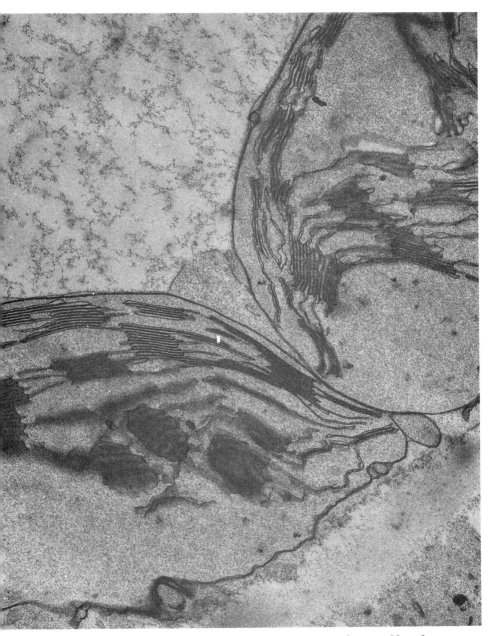

FIGURE 4. A thin section of spinach cell showing two adjacent chloroplasts. KMnO₄ staining. × 53,000.

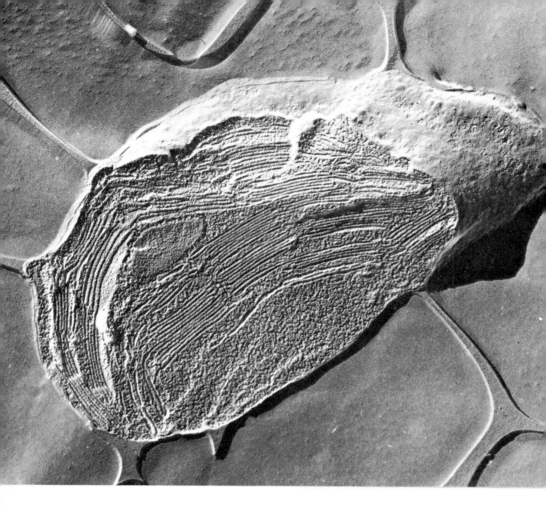

FIGURE 5. A replica of a spinach chloroplast isolated by the Jensen-Bassham technique after freeze fracturing (11).

associated particles, indicating that they differ from those in the stroma (15).

The kinds of membrane fracture faces found in such a washed preparation are shown in Figure 7. Two principal kinds are seen: the face labeled B which has particles 175Å x 90Å associated with it, and the C face which appears to possess smaller particles than the B face. There has been considerable controversy over the interpretation of these two faces. Mühlethaler et al. (13) initially maintained that the B and C faces represented the interface between the membranes and the frozen aqueous medium in which the membranes were suspended. However, a number of experiments argued against this interpretation. One of the most graphic is indicated in Figures 8 and 9 (16). If the

FIGURE 6. A replica of spinach chloroplast isolated by a cruder technique after freeze fracturing. $\times$ 72,000.

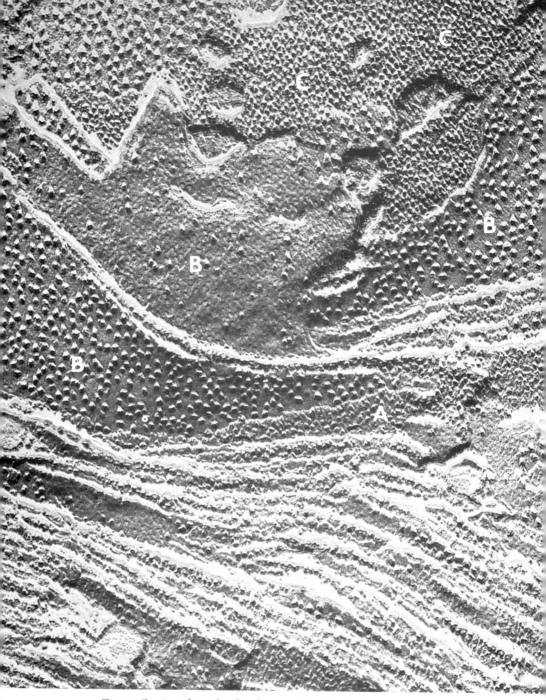

FIGURE 7. A replica of isolated spinach thylakoids showing the various fracture faces A, B, and C. $\times$ 130,000.

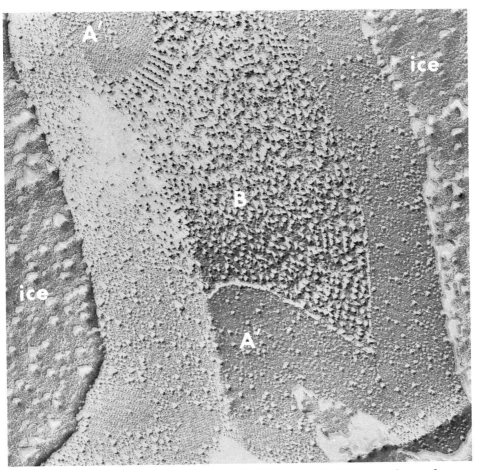

FIGURE 8. A replica of a spinach thylakoid showing the B fracture face and the adjacent A′ surface exposed by sublimation of ice. × 70,000. See Figure 10 for explanation.

Mühlethaler view was correct, then sublimation of ice from the specimen following fracturing (deep etching) should reveal surfaces resembling the fracture face. Figures 8 and 9 show this is not the case. The surfaces exposed by subliming ice adjacent to B and C fracture faces are very different in appearance from the fracture faces themselves. Data such as these suggest that thylakoids and other membranes

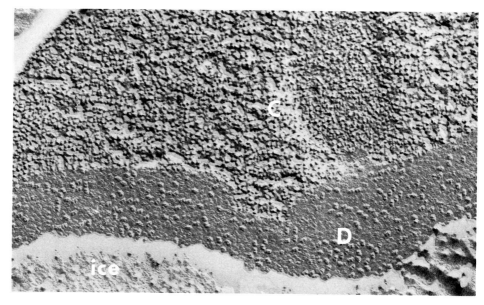

FIGURE 9.   A replica of a spinach thylakoid showing the C fracture face and the adjacent D surface exposed by sublimation of ice. ×81,000. See Figure 10 for explanation.

are actually split during freeze fracturing, much like the opening of a sandwich. An interpretation of these results is given in Figure 10.

We have suggested that the membrane-associated fracture planes which occur during freeze etching arise from splits along the hydrophobic regions of the structure. Our reasoning is briefly that it is primarily hydrophobic forces which hold membranes together in aqueous medium. Once the restoring forces of liquid water are removed by freezing, the weakest bonding in the membrane probably occurs in the hydrophobic regions. Deamer and Branton (7) have done beautiful model experiments to support this point of view. If the splitting hypothesis is correct, the particles seen in great relief on the B fracture face and in less relief on the C face are in reality buried within the intact membrane.

What is the chemical nature of these particles and the then embedding membrane matrix? The total thylakoid is about 50 percent lipid and 50 percent protein. Enzymatic degradation and antibody experiments have indicated that both lipid and protein are present at the membrane surface. It is also known from X-ray diffraction evidence

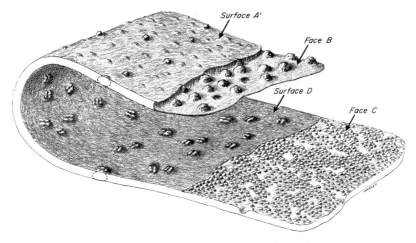

Surface A'

Face B

Surface D

Face C

FIGURE 10.   A diagrammatic representation of the relative locations of the various thylakoid surfaces and fracture faces seen by freeze etching.

and birefringence studies that the thylakoid has a much more complex organization than a simple membrane such as myelin (4). What then is the chemical nature of the particles and embedding matrix revealed by freeze etching? Partial answers to this question are suggested by several kinds of observation. Intact thylakoids can be spotted on a carbon film and then extracted with solvents which remove up to 90 percent of the lipid. The protein residue can then be shadowed with heavy metal and observed in the electron microscope. Such a preparation is seen in Figure 11 (14). A subunit structure similar in appearance to the B face particles remains. The relationship of these particles to those on the B face would be very tenuous if it were not for the fact that the particle periodicities in linear arrays are very similar in the two preparations. This suggests in a tentative way that the B face particles are protein associations embedded in a lipid matrix. The embedding matrix may be both lipid and protein with polar regions of these molecules facing the external aqueous medium. The surfaces of the particles remain exposed, creating a surface mosaic of lipid and protein and accounting for the susceptibility of thylakoids to damage by both proteases and lipases. (3).

This leaves us with the following question: What function do the B face particles perform? This is not an easy question to answer because the proteins of thylakoids appear to be largely hydrophobic in

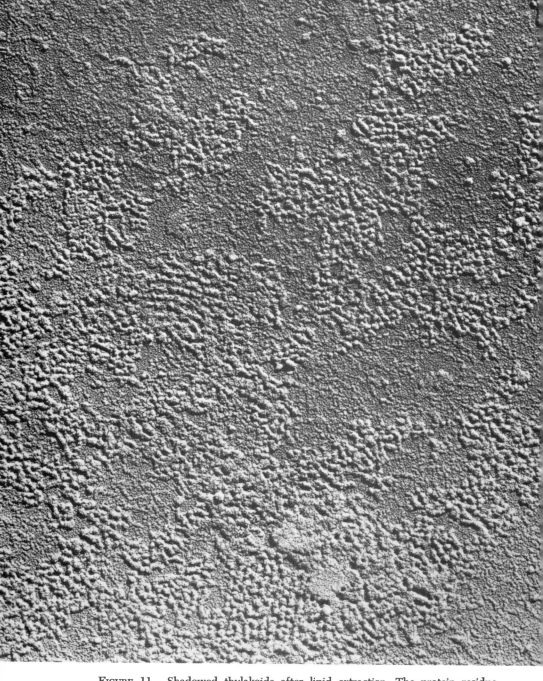

FIGURE 11. Shadowed thylakoids after lipid extraction. The protein residue remains showing particle periodicities which correspond to those seen in B fracture faces. × 138,000.

character and thus far are solubilized only in the presence of a variety of detergents. High concentrations of strong detergents such as SDS (sodium dodecylsulfate) are particularly disruptive. The proteins extracted by such treatments are much smaller than the B face units whose function we would like to determine. One possibility for isolating these units intact is to extract them from thylakoids previously treated with protein crosslinking agents such as glutaraldehyde. One might expect that here the original association of the hydrophobic protein units could be preserved during the course of detergent extraction (9).

We suggested some years ago that the B face units with their surrounding matrix might be functional units of electron-transport and gave these morphological units the name "quantasome." The split membranes which reveal these units in much greater detail are not the quantasomes as originally defined with their embedding matrices, but are quantasome cores (16). They are apparently protein in nature and must eventually be reconciled with thylakoid function. Thus far it seems there are about 55 chlorophylls coupled to a rate-limiting reaction site in system II reactions and about 450 coupled to rate-limiting sites in system I reactions (12). Thylakoid density and chemical composition suggest that the B face particle (the quantasome core) with its embedding matrix is of sufficient size to include the main electron transport functions of photosynthesis. This suggestion has been both challenged and supported over the years. It is hoped that our technical abilities in thylakoid fractionation and reconstitution will improve to the point that we can eventually give a definitive answer to thylakoid particle function.

Literature Cited

1. Arnon, D. I., M. B. Allen, and F. R. Whatley. 1954. Photosynthesis by isolated chloroplasts. Nature, 174:394-396.

2. Arnon, D. I., H. Y. Tsujimoto, and B. D. McSwain. 1965. Photosynthetic phosphorylation and electron transport. Nature, 207:1367-1372.

3. Bamberger, E. S., and R. B. Park. 1966. Effect of hydrolytic enzymes on the photosynthetic efficiency and morphology of chloroplasts. Plant Physiol., 41:1591-1600.

4. Branton, D. 1968. Structure of the photosynthetic apparatus. Photophysiol., 3:197-224.

5. Branton, D., and R. B. Park. 1967. Subunits in chloroplast lamellae. J. Ultrastruct. Res., 19:283-303.

6. Cloud, W. 1967. Mystery of the green bodies. Senior Science, 21:9-11.

7. Deamer, D., and D. Branton. 1967. Fracture planes in an ice-bilayer model membrane system. Science, *158*:655-657.

8. Engelmann, T. W. 1894. Die Erscheinungsweise der Sauerstoffausscheidung chromophyllhaltiger Zellen im Licht bei Anwendung der Bacterienmethode. Arch. Ges. Physio., (Pflügers) *57*:375-386.

9. Hallier, U. W., and R. B. Park. 1969. Photosynthetic light reactions in chemically fixed thylakoids. Plant. Physiol., *44*:544-546.

10. Hatch, M. D., C. R. Slack, and H. S. Johnson. 1967. Further studies on a new pathway of photosynthetic carbon dioxide fixation in sugar-cane and its occurrence in other plant species. Biochem. J., *102*:417-422.

11. Jensen, R. G., and J. A. Bassham. 1966. Photosynthesis by isolated chloroplasts. Proc. Nat. Acad. Sci. U. S., *56*:1095-1101.

12. Kelly, J., and K. Sauer. 1968. Functional photosynthetic units sizes for each of the two light reactions in spinach chloroplasts. Biochem., *7*:882-890.

13. Mühlethaler, K., H. Moor, and J. W. Sarkowski. 1965. The ultrastructure of chloroplast lamellae. Planta (Berlin), *67*:305-323.

14. Park, R. B. 1965. Substructure of chloroplast lamellae. J. Cell Biol., *27*:151-161.

15. Park, R. B., and A. O. Pfeifhofer. 1969. The effect of EDTA washing on the structure of spinach thylakoids. J. Cell Sci., *5*:313-319.

16. Park, R. B., and A. O. Pfeifhofer. 1969. Ultrastructural observations on deep etched thylakoids. J. Cell Sci., *5*:299-311.

17. Park, R. B., and N. G. Pon. 1961. Correlation of structure with function in *Spinacea oleracea* chloroplasts. J. Mol. Biol., *3*:1-10.

18. Trebst, A. V., H. Y. Tsujimoto, and D. I. Arnon. 1958. Separation of light and dark phases in the photosynthesis of isolated chloroplasts. Nature, *182*:351-355.

19. Trench, R. K., R. W. Greene and B. G. Bystrom. 1969. Chloroplasts as functional organelles in animal tissues. J. Cell Biol., *42*:404-417.

Properties and Biological Significance of Mitochondrial DNA

MARGIT M. K. NASS
Department of Therapeutic Research
University of Pennsylvania School of Medicine, Philadelphia

STUDIES OF THE STRUCTURE and function of cytoplasmic organelles, especially mitochondria and chloroplasts, have followed new directions during the past few years. It has become clear that mitochondria and chloroplasts function in the cell not merely as energy donors, but also have the capacity to control, at least partially, their own duplication.

In the early cytological literature on mitochondria, Altmann (1890) hypothesized that the mitochondria of a cell divide like bacteria and carry genetic information. Wallin (1927) contended that mitochondria of eukaryotic cells arose as endosymbiotic microorganisms during the evolution of the cell. Ephrussi (1953) described mutations in yeast that affected certain respiratory components of mitochondria. The mutations were inherited by a non-Mendelian mechanism. These experiments were not well understood at the time when mitochondrial DNA was still unknown. The early ideas suggesting mitochondrial autonomy or semi-autonomy can now be more seriously considered, based on the recent use of more sophisticated methods developed in the field of molecular biology. These methods include the combined use of genetic, electron microscopic, biochemical, and physicochemical analysis. The literature applying these techniques to the study of cytoplasmic organelles during the past few years has become too gigantic to be reviewed here in detail. (For reviews see Nass, 1969, 1967; Borst and Kroon, 1969.)

This recent surge of activity is based largely on various studies that have shown indisputably that mitochondria and chloroplasts contain organelle-specific DNA and RNA. By applying combined electron microscopical and cytochemical methods, Nass and Nass (1962, 1963) have shown that mitochondria contain DNA fibrils similar to the ones seen in bacteria. Mitochondrial DNA and RNA have since been studied extensively in many laboratories and found to be structurally distinct from the DNA and RNA occurring in other parts of the cell. Specifically, mitochondrial DNA of most higher animal cells is a circular

41

molecule of much lower molecular weight than the apparently linear DNA of the cell nucleus (Nass, 1966; Van Bruggen et al., 1966; Sinclair and Stevens, 1966). Mitochondrial DNA of several unicellular eukaryotic cells and plants contain linear rather than circular DNA. In general, mitochondrial DNA also differs in base composition from nuclear DNA (cf. Nass, 1969a; Borst and Kroon, 1969). Mitochondria also have distinct ribosomes that are smaller than cytoplasmic ribosomes, and the ribosomes, as well as the RNA extracted from them, have lower sedimentation coefficients as compared with their cytoplasmic counterparts (Küntzel and Noll, 1967; Rogers et al., 1967). Mitochondria have specific species of aminoacyl transfer RNA and their respective aminoacyl t-RNA synthetases (Brown and Novelli, 1968; Buck and Nass, 1968, 1969). Mitochondria also have the capacity to synthesize DNA, RNA, and protein (see reviews cited above). Furthermore, a specific mitochondrial DNA polymerase has been isolated and partially purified (Meyer and Simpson, 1968; Kalf and Ch'ih, 1968).

There is increasing evidence that mitochondria and chloroplasts are self-duplicating organelles, i.e., new mitochondria do not arise *de novo* but are formed from pre-existing mitochondria by division (Luck, 1963; cf. Nass, 1967). It is highly probable but not yet proven that the DNA of cytoplasmic organelles has a genetic function and at least partially controls the biosynthesis of these organelles. Some of the most compelling evidence stems from studies of respiratory mutants of *Neurospora* and yeast, which have mitochondrial DNA with an altered base composition, and from DNA-RNA hybridization studies (cf. reviews cited above).

One of the most urgent questions must be resolved: To what extent do mitochondria control their own duplication? Furthermore, we wonder: How did these organelles originate in the evolution of the cell? Some answers to these questions are beginning to emerge, essentially through studies of basic problems as outlined in the following sections.

Ultrastructure of mitochondrial DNA in situ

Various electron microscopic and biochemical studies in our laboratory during the past few years have led to the construction of a three-dimensional model of an L-cell mitochondrion (Figure 1). The organelles are filamentous and branched. Mitochondrial DNA is circular and is located in discrete areas of the mitochondrial matrix and may

FIGURE 1. Diagrammatic representation of a typical L-cell mitochondrion show-ing the possible arrangement of circular DNA within a mitochondrion. (From Nass, 1969 a.)

be attached to portions of the membranes. Whether the DNA molecules *in vivo* are coiled as shown in the diagram is not yet resolved. The following discussion presents some of the data that have led to this model of a mitochondrion.

In ultrathin sections of mitochondria from many cell types, the DNA may be visualized by electron microscopy in discrete areas of low density in the mitochondrial matrix (Nass and Nass, 1963; Nass et al., 1965). Depending on the fixation method used, DNA appears either as thin fibrils 20 Å in width after fixation with OsO_4 followed by uranyl acetate (Figure 2) or in a clumped form if uranyl acetate is omitted (Figure 3). The DNA of the nucleus does not form fibrils under these conditions. The fibrils can be specifically digested by deoxyribonuclease. They are frequently found attached to the mito-chondrial membrane. The DNA of prokaryotic microorganisms has fixation characteristics similar to mitochondrial DNA, which may indi-cate that mitochondrial and prokaryotic DNA is not bound to histones as is nuclear DNA. The clumped appearance of the DNA is illustrated in Figure 4 for *Mycoplasma*, a small bacteria-like organism which lacks a cell wall. The mitochondrion in Figure 3 also shows small electron-dense particles which are smaller than cytoplasmic ribosomes and presumably represent mitochondrial ribosomes. The DNA of the mitochondrion shown occupies two discrete areas. Studies using serial sections of mitochondria have shown that these filamentous and often branched organelles not only vary in length, but also in the number of DNA-containing regions or nucleoids, usually two to three, but some-times up to six. Multiple nucleoids are also common in bacteria. The

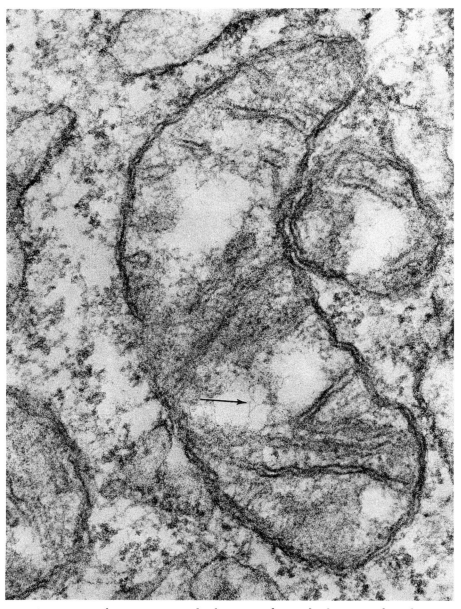

FIGURE 2. Electron micrograph showing a longitudinal section through a mitochondrion of L-cells, fixed with buffered 2% osmium tetroxide followed by 0.5% aqueous uranyl acetate. Arrow points to DNA fiber of 20 Å diameter, partially visible in an area of low matrix density. × 90,000.

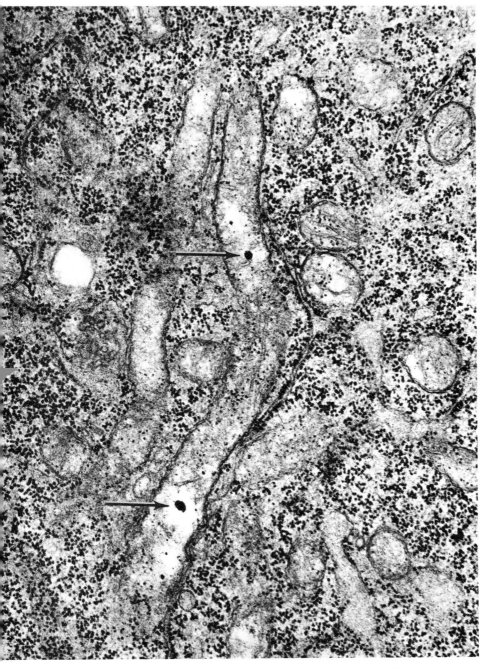

FIGURE 3. Electron micrograph showing part of a filamentous mitochondrion (center), sections through other mitochondria, and ribosomes. Mitochondrion in

center reveals two nucleoids or DNA-containing areas (arrows). The DNA is preserved in a clumped form. Fixation was 2.5% glutaraldehyde in 0.1 M buffer, followed by 2% OsO₄ in the same buffer. Cytoplasmic ribosomes are well preserved; the smaller granules within profiles of mitochondria are presumably mitochondrial ribosomes. The particles are not seen in other areas, such as vesicles of the endoplasmic reticulum or outside the cell (upper left corner). × 45,000.

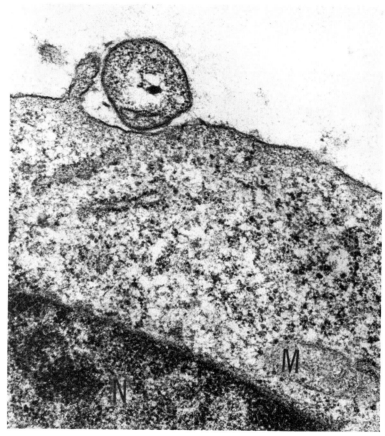

FIGURE 4. Electron micrograph of the microorganism *Mycoplasma* at the cell surface of a cultured Chang liver cell. The DNA of this organism (arrow), as in mitochondria, is preserved in a clumped form after fixation in buffered OsO₄. M, mitochondrion; N, nucleus. × 40,000.

electron micrographs, however, do not reveal how many molecules each nucleoid contains. Another technique was therefore applied by which whole DNA molecules can be visualized in the electron microscope as they are liberated from isolated mitochondria by osmotic shock. Figure 5 shows a bacterium that has been subjected to this procedure. The organisms, suspended in 4 M ammonium acetate, are osmotically lysed on a monolayer film of protein spread on the surface of water. The membranes rupture and the DNA is extruded and adsorbed to the protein film. The molecules are visualized by shadowing from all sides with vaporized metal. The technique avoids extensive handling of the

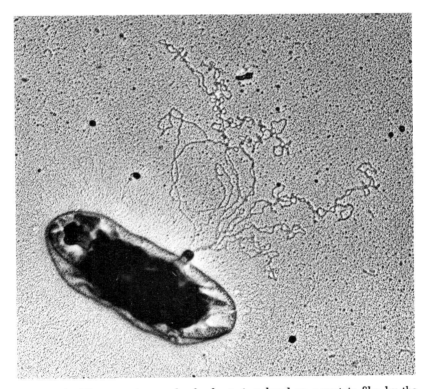

FIGURE 5.   Electron micrograph of a bacterium lysed on a protein film by the osmotic shock technique. Part of the DNA is extruded through an opening in the cell wall. The DNA appears intact; there are no free ends visible. Part of the DNA molecule is supercoiled. × 35,000.

DNA and thereby preserves it relatively intact. Figure 6 shows an osmotically lysed L-cell mitchondrion. Circular DNA molecules are generally seen folded in half; the halves are either coiled around each other or remain closely associated after unwinding. The DNA is frequently attached to fragments of the mitochondrial membranes. Most DNA molecules measure about 5 μ in circumference, which corresponds to a molecular weight of about 10 x 10^6 daltons. Usually up to six DNA molecules have been found associated with a ruptured mitochondrion. This agrees well with the chemically determined DNA content of an L-cell mitochondrion, which is 53 x 10^6 daltons or an average of 5 DNA molecules of monomer molecular weight (Nass, 1966).

In addition to DNA monomers, 2 to 4 percent of the DNA of osmotically lysed mitochondria consisted of double-size circular molecules or dimers (Nass, 1969b). The contour lengths of various forms of DNA molecules are shown in Table 1. A dimer was sometimes attached to another dimer or monomer by a common knob-like central point (Figure 7). The fact that these DNA molecules are connected

Table 1. MEASUREMENTS OF THE LENGTH OF MITOCHONDRIAL DNA EXTRUDED BY OSMOTIC SHOCK (From Nass, 1966)

Preparation	No. of observations (molecules or aggregates of molecules)	Total length (μ)	Molecular weight	Length (total) — Length (single)
Single twisted or open circles	45	5.21 ± 0.20	10.0 x 10^6	1.0
Double size twisted or open circles	11	10.1 ± 0.4	19.6 x 10^6	2.0
Multiple forms resolved as linked or overlapping single (1) and double (2) sizes				
4 x (1)	3	21.3 ± 1.0		4.1
2 x (1) +2 x (2)	2	32.8 ± 1.3		6.3
Tangled DNA, not resolvable into individual molecules..	9	10.7 ± 1.2		2.0
	9	15.1 ± 0.8		2.9
	6	21.7 ± 0.9		4.1
	1	31.5		6.0

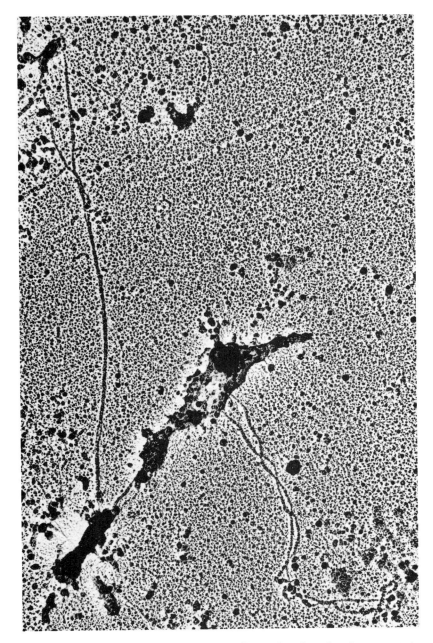

FIGURE 6. Electron micrograph of an L-cell mitochondrion lysed on a protein film by the osmotic shock technique. Mitochondrion is fragmented into several pieces of membrane, and parts of two circular DNA molecules are visible on left and lower right. × 49,000. (From Nass, 1967.)

at a central point suggests that they are derived from one mitochondrial nucleoid.

The studies with L-cells thus far have shown that the number of DNA molecules per mitochondrion may be variable. Also the number of nucleoids per mitochondrion may vary, as well as the number of DNA molecules contained in a nucleoid, allowing for the presence of up to six molecules per organelle. These DNA molecules may be mixtures of monomers and dimers that are either free or attached to each other.

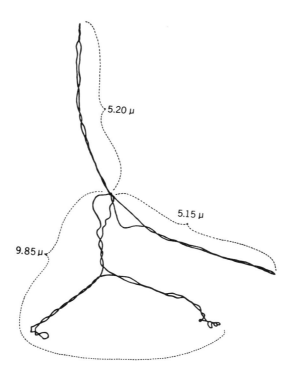

FIGURE 7.   Diagram of circular DNA molecules extruded by osmotic shock from L-cell mitochondrion. One unicircular dimer and two monomers (or an interlocked dimer) are attached at a central point. These molecules may represent the content of one mitochondrial nucleoid. (From Nass, 1969 b.)

Properties of Isolated Mitochondrial DNA

Mitochondrial DNA generally constitutes only 0.1 to 1 percent of total cellular DNA. There are several methods by which mitochondrial DNA can be separated from contaminating nuclear DNA. Mitochondrial and nuclear DNA of many organisms differ in average base composition and may thus be isolated as separate bands of different buoyant density after equilibrium centrifugation in cesium chloride gradients. However, such differences may be very small for mitchondrial and nuclear DNA from many types of mammalian cells. Mitochondrial and nuclear DNA's from mouse L-cells, for example, have buoyant densities of 1.698 and 1.703 g cm^{-3}, corresponding to a guanine plus cytosine content of 39 and 43 percent (Nass, 1968, 1969 c). The difference is not large enough to allow separation of the fractions on gradients. In this case, contaminating nuclear DNA can be removed by a brief treatment of the isolated mitochondria with deoxyribonuclease which does not penetrate into structurally intact organelles. Mitochondrial DNA of a typical preparation, when examined in the electron microscope, consists of a mixture of highly twisted circular forms and loosely twisted or open types (Figure 8). The twisted or supercoiled structure is typical of covalently closed DNA, which is free of single-strand breaks, and the loosely twisted forms may represent DNA with one or more single-strand scissions.

The contour lengths of circular DNA monomers from mitochondria of human, avian, amphibian, and sea urchin cells are in the range of 4.5 to 5μ. Although some variation is undoubtedly due to technical factors in the hands of different investigators, true size differences do exist. The size measurements obtained from a mixture of DNA of mouse L-cells and chicken liver mitochondria followed a bimodal distribution with peak categories (4.7-4.8μ and 5.1-5.2μ, respectively) corresponding to the size of each DNA when spread individually (Nass, 1969 c).

The circular structure of mitochondrial DNA does not appear to be universal. The bulk of yeast mitochondrial DNA has been reported to consist of linear filaments and some circular molecules about 4.5 to 5μ in length. Mitochondrial DNA's of many eukaryotic microbial and plant cells appear to be linear also, but have a higher molecular weight than that corresponding to the 5μ circles (see review papers cited above).

A very useful preparative technique has been described by Radloff and others (1967) by which DNA molecules can be separated on cesium chloride-ethidium bromide gradients on the basis of molecular topology rather than base composition. The fluorescent dye

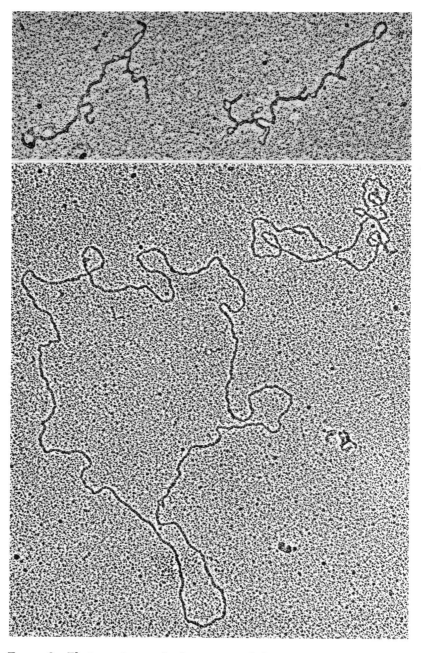

FIGURE 8. Electron micrograph of two supercoiled monomeric DNA molecules (top), a loosely coiled DNA monomer (right), and an open unicircular dimer (left), isolated from L-cell mitochondria. × 40,000.

ethidium bromide intercalates between the stacked base pairs of the helical double-stranded DNA molecules. If dye binds to a covalently closed, twisted ring molecule, the unwinding of Watson-Crick helical turns is compensated by a corresponding change in the number of tertiary twists or supercoils so that the total number of turns in the closed molecule remains the same. At low dye concentrations the number of originally right-handed superhelical turns is reduced to zero, and as more dye is intercalated at high concentrations, superhelices of opposite handedness are introduced. As a result of the changes in molecular conformation and in the free energy of the dye-binding reaction, the maximum amount of dye that can be bound by covalently closed molecules is actually smaller than the amount bound by linear or nicked circular molecules. Furthermore, the more dye is inserted into a DNA molecule the lower becomes its buoyant density in cesium chloride. To isolate and separate covalently closed circular from nicked circular and linear DNA, DNA is centrifuged to equilibrium in a solution of cesium chloride and ethidium bromide as shown in Figure 9. Examination of the gradients in ultraviolet light may reveal three fluorescent bands. The least dense upper band contains a mixture of nicked circles and linear DNA, and the densest lower, usually more faintly fluorescent band, contains exclusively covalently closed circles that may vary in size. A minor third band of intermediate density is enriched with circular DNA molecules forming interlocked circles of two or more basic size units (Figure 9b). The double size units or dimers are most frequent. The interlocked dimers that are found in the middle band may be of intermediate density because they have single-strand nicks in one submolecule and are covalently closed in the other. In some cell types, the lowest band, containing only covalently closed circles, also includes about 10 percent unicircular dimers and some interlocked dimers (Figure 10). Leukocytes derived from leukemic patients (Clayton and Vinograd, 1967) and mouse L-cells (Nass, 1969, b, c,), which are malignant cells, are particularly enriched in the unicircular dimers.

The physicochemical properties of covalently closed and nicked circular forms of mitochondrial DNA have recently been studied extensively and found to conform with most of the criteria that have been established to describe the properties of intact and nicked circular DNA of polyoma viruses (Vinograd and Lebowitz, 1966). In summary, the twisted supercoiled circular DNA, which predominates in fresh preparations of circular viral or mitochondrial DNA, is covalently closed (component I) and the open or loosely twisted circular form has one or more single-strand scissions (component II). The former can be converted to the latter by the introduction of single-strand breaks, for example, by mechanical shearing or by treatment with very low con-

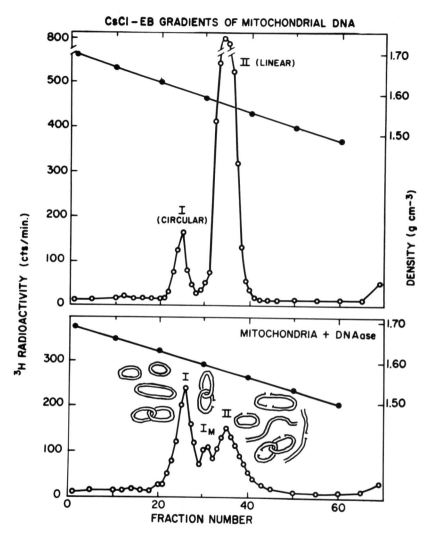

FIGURE 9. Mitochondrial DNA centrifuged to equilibrium in a gradient of cesium chloride-ethidium bromide. (a) Mitochondria were not pretreated with deoxyribonuclease to remove contaminating nuclear DNA; (b) mitochondria received treatment with deoxyribonuclease. A few of the typical forms of DNA found in the fractions are shown diagrammatically. Covalently closed DNA monomers and dimers are concentrated in band I, various DNA forms having single-strand scissions are found in the other bands. (Modified from Nass, 1969 c.)

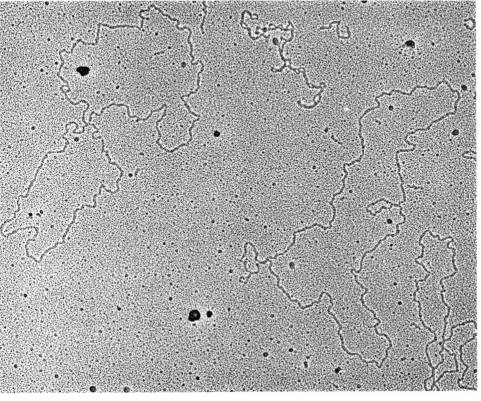

FIGURE 10.   Electron micrograph of two types of circular covalently closed dimers and a supercoiled monomer (upper center). × 30,000.

centrations of the DNA degradation enzyme deoxyribonuclease. The single-strand breaks allow the molecule to rotate around the bond opposite the break and thus release the supercoils. If both DNA strands are cut at opposite points, linear molecules are formed, which are frequently referred to as component III. The properties of component I differ from those of II and III due to the topological restraint to unwinding of the double-stranded helix, imposed by the covalently closed structure. Consequently, component I, as opposed to II or III, greatly resists denaturation or strand separation by heat and alkali and renatures spontaneously to the original state upon rapid cooling or neutralization because both strands remain aligned during denaturation and "zip" together as hydrogen bonds reform. The two strands of nicked circles and linear molecules, on the other hand, become completely separated during denaturation to form single-stranded circles and single-stranded linear fragments. The hydrodynamically more compact component I also has a higher sedimentation velocity in cesium chloride or sucrose gradients at neutral pH, and, more so, at alkaline pH above about pH 12.5. This form of DNA has a higher buoyant density in alkaline cesium chloride solutions. Furthermore, as discussed earlier, component I binds less of the intercalating dye ethidium bromide at high dye concentrations than nicked circular and linear DNA, and therefore bands at a higher density level in cesium chloride-ethidium bromide gradients. Experiments related to these properties and electron micrographs of double-and single-stranded mitochondrial DNA molecules following denaturation have been presented (Nass, 1969 c).

Informational Content of Mitochondrial DNA

To what extent do mitochondria determine their own duplication? Are all mitochondrial macromolecular components coded for by mitochondrial DNA or are some under the control of nuclear genes? Is the informational content limited to one molecule of DNA or do additional molecules of DNA in the same organelles duplicate or increase the genetic information?

It can be inferred from many studies that mitochondrial DNA has genetic functions (cf. Nass, 1969 a). Nevertheless, this short piece of DNA is probably insufficient to code for all the macromolecular components of a mitochondrion.

A 5μ circular molecule has a molecular weight of 9-10 x 10^6 daltons (about 16,000 base pairs) and can code for only about 5,000 amino acids or 30 proteins of molecular weight 20,000. There is now good evidence that the nucleus codes for some mitochondrial components, such as cytochrome c (Kadenbach, 1968; Freeman, et al., 1967). The idea that a mitochondrial DNA *monomer* contains the basic information, rather than that mitochondria contain several genetically different DNA molecules, is presently favored. Our studies, described earlier, show that variable numbers of DNA monomers and dimers are found in different mitochondria as well as in individual nucleoids. This implies redundancy of informational content. Furthermore, Borst and Kroon (cf. 1969) showed by renaturation kinetics that mitochondrial DNA, like bacterial and viral DNA, is highly homogeneous in base sequences. Nevertheless, studies are under way to test by combined electron microscopical and biochemical methods whether all molecules in a DNA population isolated from mitochondria are identical (Nass, 1968). The methods involve heating the DNA in formaldehyde, which produces local points of strand separation because adenine-thymine-rich regions melt out preferentially (Figure 11, arrows). The location of these sites appears to be quite reproducible on different molecules, suggestive of homogeneous DNA.

Present evidence strongly suggests that mitochondrial DNA codes for the synthesis of the inner mitochondrial membrane and for mitochondria-specific RNA, such as ribosomes and transfer RNA (cf. reviews cited earlier). Studies in our laboratory have shown that mitochondria of rat liver contain many organelle-specific species of aminoacyl transfer RNA (Buck and Nass, 1968, 1969). These were tested to determine whether they hydridize, i.e., have complementary base sequences, with mitochondrial DNA. Leucyl-tRNA was tested first. Chromatographically, mitochondrial leucyl-tRNA contains RNA species which are not found in cytoplasmic leucyl-tRNA. When the aminoacyl-t RNA synthetases are interchanged, the cytoplasmic enzyme does not acylate the mitochondria-specific species of leucyl-tRNA. In hybridization studies we have shown that mitchondrial leucyl-tRNA hybridizes significantly with mitochondrial DNA in contrast to cytoplasmic leucyl-tRNA which reacts negligibly (Nass and Buck, 1969). Figure 12 illustrates a competition experiment where increasing

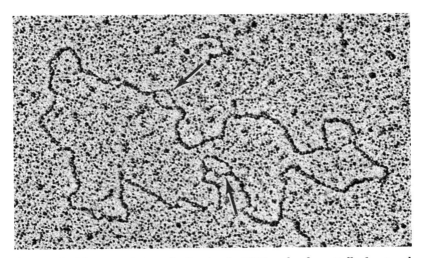

Figure 11.   Electron micrograph of a circular DNA molecule partially denatured by heating in formaldehyde. Arrows point at two sites of single-strand separation. Molecule has shortened after this procedure. × 100,000. (From Nass, 1968.)

amounts of unlabeled aminoacyl-tRNA from mitochondria, from extra-mitochondrial cytoplasm, and from the bacterium *E. coli.* compete with labeled mitochondrial leucyl-tRNA for sites on mitochondrial DNA that have complementary base sequences. Mitochondrial tRNA competes best, *E. coli* tRNA not at all, and cytoplasmic tRNA very slightly. The latter reaction is probably due to the presence of one species of leucyl tRNA which mitochondria and the cytoplasm have in common. These and studies with other aminoacyl-tRNA's show that mitochondrial and cytoplasmic tRNA's differ in primary base sequences and that mitochondrial DNA has the capacity to code for organelle-specific tRNA. Mitochondrial ribosomal RNA has also been shown to hybri-

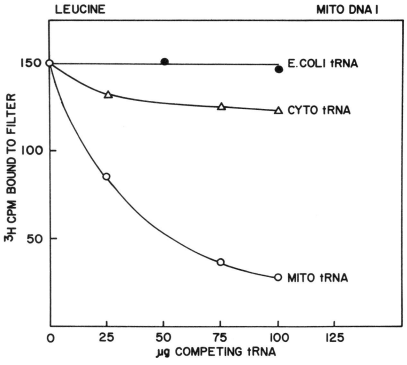

FIGURE 12. Hybridization of mitochondrial leucyl-tRNA with mitochondrial DNA of rat liver. Competition between mitochondrial ³H-leucyl tRNA and increasing amounts of unlabeled aminoacyl tRNA for hybridization sites on mitochondrial DNA. o - o, mitochondrial tRNA; △ - △ cytoplasmic tRNA; ● - ● *E. coli* tRNA. (From Nass and Buck, 1969.)

dize with mitochondrial DNA (Wintersberger and Viehhauser, 1968).
Theoretically, approximately 30 percent of a 5μ DNA circle would be
required to be complementary to mitochondrial ribosomal RNA and
10 percent to code for 20 species of tRNA of molecular weight 25,000.
However, the exact number of organelle-specific species of tRNA is
not yet known. Mitochondrial DNA appears to be large enough to
code for tRNA and ribosomal RNA, but it is apparently not large
enough to code for all the aminoacyl-tRNA synthetases. These and
other enzymes, e.g., dehyrogenases, methylases, DNA and RNA
polymerases, as well as the outer mitochondrial membrane may be
coded for by the nucleus. It should also be interesting to determine
whether the high molecular-weight linear DNA from mitochondria of
lower eukaryotic cells and plants contain more information than the
5μ circle or simply a higher degree of redundancy of information.

On the Degree of Independence of Cytoplasmic Organelles

Although isolated cytoplasmic organelles, such as mitochondria and
chloroplasts, are capable of synthesizing DNA, RNA, and protein,
there is as yet no convincing demonstration that they can replicate
in vitro. The old idea is presently revived that in the evolution of a
eukaryotic cell mitochondria and chloroplasts arose from free-living
organisms that were ingested by a primitive cell. If this did indeed
happen, the organelles of today have obviously lost some of the original
functions and acquired others in response to the cell's needs. If the
nucleus now codes for some mitochondrial components, it would undoubtedly be difficult to propagate isolated mitochondria in culture unless one succeeded in finding the right, obviously complex formula of
supplements that possibly include messenger RNA from the nucleus
and a cytoplasmic protein synthesis system.

We have been interested in the problem of the degree of inde-
pendence which may be achieved by isolated cytoplasmic organelles
that contain DNA. We have therefore attempted to introduce isolated
heterospecific cytoplasmic organelles into mammalian cells. Advantage
was taken of the phagocytic properties of mouse fibroblasts (L-cells)
grown in suspension culture. These animal cells incorporated isolated
green chloroplasts of spinach and African violets, as well as isolated
mitochondria of chicken liver (Nass, 1969d). The organelles resided in
the cytoplasm and were not contained in vacuoles or digestion vesicles
(Figure 13). The green animal cells divided like normal cells. Chloroplasts were followed for five cell generations or five days, at which
time hybrid cells were greatly outnumbered by nongreen progeny cells.
The ingested chloroplasts retained their structural integrity as de-

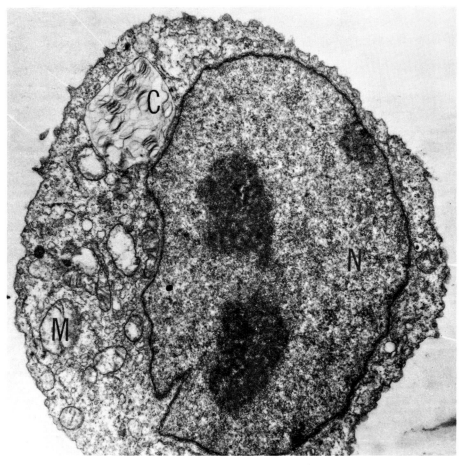

FIGURE 13. Electron micrograph of mouse L-cell that has phagocytically incorporated isolated chloroplasts of spinach. Electron micrograph taken after 24
hours in culture. Ch, chloroplast; M, mitochondria; N, nucleus. × 40,000. (From
Nass, 1969 d.)

termined by electron microscopy of the organelles and cells, by analysis
of photochemical activity (light-stimulated Hill reaction activity and
CO_2 fixation), and by analysis of the DNA in chloroplasts reisolated
from cells after one to two days in culture. Functional chloroplasts have
been reported to occur naturally as symbiotic organelles in the digestive
gland of the mollusc *Elysia* (Taylor, 1968). Our studies indicate that

animal cells can indeed incorporate chloroplasts and mitochondria. One may also visualize that primitive cells may have engaged in such a process. The degree to which these ingested heterospecific organelles can function and multiply in the animal cells under various conditions of culture is currently under investigation.

Conclusion

The task which lies ahead will continue to involve (1) the identification and properties of specific mitochondrial components and (2) the synthesis, functional interrelation, and control of these components. The control appears to be dual, involving mitochondrial as well as nuclear messages. It remains to be determined exactly to what degree mitochondria control their own duplication in different cell types. It will be interesting to see whether mitochondria with high molecular-weight linear DNA in cells at a low evolutionary level are more independent and control more of their own functions than mitochondria with 5μ DNA in higher cells. The genetic function of mitochondrial DNA undoubtedly will become better understood in eukaryotic microorganisms, such as yeast, *Neurospora,* or *Tetrahymena,* where mutants are available, than in cells of higher vertebrates where mutations may be lethal or difficult to detect and utilize as a tool.

ACKNOWLEDGMENTS: This work has been supported by NIH grant 5-POI-AI 07005 and Career Development Award I-K03-AI-0803 from the U. S. Public Health Service. I am grateful to Mrs. A. Hathaway, Mr. W. Colquhoun, and Mr. J. R. W. Hobbs for skillful technical assistance.

Literature Cited

Altmann, R. 1890. Die Elementarorganismen und ihre Beziehungen zu den Zellen. Leipzig: Veit Co.

Borst, P., and A. M. Kroon. 1969. Mitochondrial DNA: physicochemical properties, replication and genetic function. Internat. Rev. Cytol., *26*:107-190.

Brown, D. H., and G. D. Novelli. 1968. Chromatographic differences between the cytoplasmic and mitochondrial tRNAs of *Neurospora crassa.* Biochem. Biophys. Res. Commun., *31*:262-266.

van Bruggen, E. F. J., P. Borst, G. J. C. M. Ruttenberg, M. Gruber, and A. M. Kroon. 1966. Circular mitochondrial DNA. Biochem. Biophys. Acta, *119*: 437-439.

Buck, C. A., and M. M. K. Nass. 1968. Differences between mitochondrial and cytoplasmic transfer RNA and aminoacyl transfer RNA synthetases from rat liver. Proc. Nat. Acad. Sci., *60*:1045-1052.

Buck, C. A., and M. M. K. Nass. 1969. Studies on mitochondrial tRNA from animal cells. I. A comparison of mitochondrial and cytoplasmic tRNA and aminoacyl tRNA synthetases. J. Mol. Biol., *41*:67-82.

Clayton, D. A., and J. Vinograd. 1967. Circular dimer and catenane forms of mitochondrial DNA in human leukemic leucocytes. Nature, *216*:652-657.

Ephrussi, B. 1953. *Nucleocytoplasmic Relations in Micro-organisms.* Oxford: Clarendon Press.

Freeman, K. B., D. Haldar, and T. S. Work. 1967. The morphological site of synthesis of cytochrome c in mammalian cells (Krebs cells). Biochem. J., *105*:947-952.

Kadenbach, B. 1968. Transfer of proteins from microsomes to mitochondria. Biosynthesis of cytochrome c. In *Biochemical Aspects of the Biogenesis of Mitochondria,* E. C. Slater, J. M. Tager, S. Papa, and E. Quagliariello, eds. Adriatica Editrice, Bari, pp. 415-429.

Kalf, G. F., and J. J. Ch'ih. 1968. Purification and properties of deoxyribonucleic acid polymerase from rat liver mitocondria. J. Biol. Chem., *243*:4904-4916.

Küntzel, H., and H. Noll. 1967. Mitochondrial and cytoplasmic polysomes from *Neurospora crassa.* Nature, *215*:1340-1345.

Luck, D. J. L. 1963. Formation of mitochondria in *Neurospora crassa.* J. Cell Biol., *16*:483-499.

Meyer, R. R., and M. V. Simpson. 1968. DNA biosynthesis in mitochondria: partial purification of a distinct DNA polymerase from isolated rat liver mitochondria. Proc. Nat. Acad. Sci., *61*:130-137.

Nass, M. M. K. 1966. The circularity of mitochondrial DNA. Proc. Nat. Acad. Sci., *56*:1215-1222.

Nass. M. M. K. 1967. Circularity and other properties of mitochondrial DNA of animal cells. In *Organizational Biosynthesis,* H. J. Vogel, J. O. Lampen, and V. Bryson, eds. New York: Academic Press, pp. 503-522.

Nass, M. M. K. 1968. Properties of organelle-associated and isolated mitochondrial DNA. In *Biochemical Aspects of the Biogenesis of Mitochondria,* E. C. Slater, J. M. Tager, S. Papa, and E. Quagliariello, eds. Adriatica Editrice, Bari, pp. 27-50.

Nass, M. M. K. 1969a. Mitochondrial DNA: Advances, problems and goals. Science, *165*:25-35.

Nass, M. M. K. 1969b. Mitochondrial DNA. I. Intramitochondrial distribution and structural relations of single- and double-length circular DNA. J. Mol. Biol., *42*:521-528.

Nass, M. M. K. 1969c. Mitochondrial DNA. II. Structure and physicochemical properties of isolated DNA. J. Mol. Biol., *42*:529-545.

Nass, M. M. K. 1969d. The uptake of isolated chloroplasts by mammalian cells. Science, *165*:1128-1131.

Nass, M. M. K., and C. A. Buck. 1969. Comparative hybridization of mitochondrial and cytoplasmic aminoacyl transfer RNA with mitochondrial DNA from rat liver. Proc. Nat. Acad Sci., *62*:506-513.

Nass, M. M. K., and S. Nass. 1962. Fibrous structures within the matrix of developing chick embryo mitochondria. Exp. Cell Res., *26*:424-427.

Nass, M. M. K., and S. Nass. 1963. Intramitochondrial fibers with DNA characteristics. I. Fixation and electron staining reactions. J. Cell Biol., *19*:593-611.

Nass, S., and M. M. K. Nass. 1963. Intramitochondrial fibers with DNA characteristics. II. Enzymatic and other hydrolytic treatments. J. Cell Biol., *19*:613-629.

Nass, M. M. K., S. Nass, and B. A. Afzelius. 1965. The general occurrence of mitochondrial DNA. Exp. Cell Res., *37*:516-539.

Radloff, R., W. Bauer, and J. Vinograd. 1967. A dye-buoyant density method for the detection and isolation of closed circular duplex DNA: the closed circular DNA in HeLa cells. Proc. Nat. Acad. Sci., *57*:1514-1521.

Rogers, P. J., B. N. Preston, E. B. Titchener, and A. W. Linnane. 1967. Differences between the sedimentation characteristics of the ribonucleic acids prepared from yeast cytoplasmic ribosomes and mitochondria. Biochem. Biophys. Res. Commun., *27*:405-411.

Sinclair, J. H., and B. J. Stevens. 1966. Circular DNA filaments from mouse mitochondria. Proc. Nat. Acad. Sci., *56*:508-514.

Taylor, D. L. 1968. Chloroplasts as symbiotic organelles in the digestive gland of *Elysia viridis* (Gastropoda: opisthobranchia). J. mar. biol. Ass. U. K., *48*:1-15.

Vinograd, J., and J. Lebowitz. 1966. Physical and topological properties of circular DNA. J. Gen. Physiol., *49*:103.

Wallin, J. E. 1927. *Symbionticism and the Origin of Species*. Baltimore: Williams and Wilkins.

Wintersberger, E., and G. Viehhauser. 1968. Function of mitochondrial DNA in yeast. Nature, *220*:699-702.

The Photosynthetic Apparatus
of Procaryotic Organisms

G. Cohen-Bazire
Department of Bacteriology and Immunology
University of California, Berkeley

In procaryotic organisms, the cytoplasmic membrane is the only unit membrane system of the cell; it possesses functional attributes which are not associated with the cell membrane of eucaryotic cells. In addition to playing a role in active transport, the membrane and its intrusions house the machinery of respiratory electron transport and photosynthetic energy conversion. Furthermore, the procaryotic chromosome appears to be anchored to the membrane, which has been postulated to play an active role in the process of nuclear and cell division (FitzJames, 1965; Ryter, 1968). To some degree, accordingly, the procaryotic cell membrane assumes roles which in eucaryotic cells are carried out by the mitochondria, the chloroplasts, and the mitotic apparatus.

As a result of this mode of cellular organization, one cannot define the site of localization of photosynthetic processes in the cells of procaryotic organisms as precisely as one can in the cells of eucaryotic organisms.

The photosynthetic apparatus of the purple and green bacteria and of the blue-green algae (the three groups of photosynthetic procaryotes) is most commonly defined as consisting only of the structures that participate in the process of photochemical energy conversion, viz., the photosynthetic pigment system and the associated machinery of electron transport. This is of course a more restricted definition than that used in the context of a eucaryotic cell, since the chloroplast also houses the entire enzymatic machinery for the conversion of CO_2 into sugar-phosphates.

Mechanisms of photosynthesis in procaryotic organisms

Of the three photosynthetic procaryotic groups, only the blue-green algae perform photosynthesis that is mechanistically of the plant type. In the purple and green bacteria, photosynthesis is never accompanied by oxygen evolution, and in fact it occurs only under anaerobic conditions. The lack of oxygen evolution characteristic of bacterial photosynthesis reflects the absence of type II reaction centers which participate in other photosynthetic organisms in the oxidation of water. The photosynthetic bacteria are consequently dependent for the reduction of CO_2 to cell material on electron donors other than water, e.g., reduced sulfur compounds, H_2, organic compounds.

These mechanistic particularities of the photosynthetic bacteria are correlated with the possession of unique photosynthetic pigment systems which contain chlorophylls and carotenoids unlike those of any oxygen-evolving photosynthetic organisms (Pfennig, 1967). The blue-green algae on the other hand, although they do contain group-specific photopigments (phycobiliproteins and special carotenoids), share with all other oxygen-evolving phototrophs the possession of chlorophyll *a* and bicyclic carotenoids such as β-carotene. Accordingly, some marked deviations from the otherwise uniform pattern of photosynthesis do occur among procaryotes. These deviations, however, are not causally related to the relative structural simplicity of the procaryotic cell since in blue-green algae both the mechanism and the chemical machinery of photosynthesis are fundamentally similar to those of eucaryotic photosynthetic groups.

Structure and organization of the procaryotic photosynthetic apparatus

It is convenient to consider separately the characteristic features of the photosynthetic apparatus in blue-green algae, in purple bacteria, and in green bacteria since these three groups can be readily distinguished from one another by the nature of their photosynthetic pigment systems (see Table 1) and also by the structure of their photosynthetic apparatus.

Blue-green algae

The photosynthetic apparatus of blue-green algae will be discussed first. This apparatus resembles that of eucaryotic photosynthetic organisms in pigment composition and function. Of the three classes of photopigments characteristic of blue-green algae, the phycobiliproteins are not membrane-bound; they disassociate readily from the membranes that bear the other photopigments upon breakage of the cells. The chlorophyll and carotenoids are borne on a system of lamellae approximately 150 to 160 Å thick, composed of two closely apposed unit membranes. The disposition of lamellae in the cell can vary considerably from species to species. In many of the unicellular blue-green

Table 1. Photosynthetic Pigments of Procaryotic Organisms

molecular ground plan

CHLOROPHYLLS		BLUE-GREEN ALGAE	PURPLE BACTERIA		GREEN BACTERIA	
		chlorophyll-a	bacteriochlorophyll a	b	bacteriochlorophylls c	a and d
R1		$-CH=CH_2$	$-C(=O)-CH_3$	?.	$-CH(-OH)-CH_3$	$-CH(-OH)-CH_3$
3,4		—	dihydro	dihydro ?.	—	—
R2		$-C(=O)-O-CH_3$	$-C(=O)-O-CH_3$	?.	$-H$	$-H$
R3		phytyl ($C_{20}H_{39}O-$)	phytyl	?.	farnesyl ($C_{15}H_{25}O-$)	farnesyl
R4		$-H$	$-H$	?.	$-CH_3$	$-H$
CAROTENOIDS		ALICYCLIC β-carotene xanthophylls	ALIPHATIC often methoxylated spirilloxanthin series		mono- or di- ARYL-CAROTENOIDS	
PHYCOBILIPROTEINS		+	—		—	

algae, the lamellae have a predominantly cortical distribution and are arrayed more or less parallel to the cell surface (Figures 1 and 2). Although it seems probable that the lamellae originate from the cell membrane, direct connections are rarely seen in thin sections, although they have been occasionally reported (Pankratz and Bowen, 1963; Lang, 1968; Edwards et al., 1968). A very characteristic feature of the photosynthetic apparatus of blue-green algae is the rather wide and regular interlamellar spacing, which approximates 300 to 450 Å. This is strikingly evident in thin sections of cells from which the chlorophyll and carotenoids have been extracted with lipid solvents after fixation (Figure 3). This treatment completely eliminates the lamellar system, the position previously occupied by the lamellae appearing uniformly electron transparent. In a number of unicellular blue-green algae, fixation with glutaraldehyde followed by osmium fixation reveals that the interlamellar spacings are occupied by electron-opaque granules with an apparent diameter varying between 250 and 350 Å (Figures 4 and 5). These granules are at times regularly arrayed in a single layer on the outer surfaces of the lamellae. This arrangement is very similar to the intrachloroplast arrangement of lamellae and granules observed by Gantt and Conti (1966) in the unicellular red alga *Porphyridium cruentum*. Similar granules have been recently described by them in three filamentous blue-green algae (Gantt and Conti, 1969). It is probable that in both red- and blue-green algae these granules represent multimolecular aggregates of phycobiliproteins. This interpretation of the relation between the phycobiliproteins and the other components of the photosynthetic apparatus is supported by the finding that membrane fragments isolated after breakage of glutaraldehyde-fixed cells of unicellular blue-green algae still carry a substantial fraction of the phycocyanin attached to them and have a granular surface when examined in negative staining (Cohen-Bazire and Lefort, 1968).

--→

FIGURE 1. *Anacystis* sp.—Longitudinal-tangential section showing the cortical disposition of photosynthetic lamellae (L). The lamellae, separated from each other and from the cytoplasmic membrane (M) by a distance varying from 300 to 400 Å, are formed by the juxtaposition of two unit membranes of an overall thickness of 165 Å. The electron opaque granules (arrows), 300 to 350 Å in diameter, visible in the interlamellar spaces on the outer surface of the lamellae are interpreted as aggregates of phycobiliproteins.

The central region of the cell is occupied by the nucleoplasm (n) surrounded by and interspersed with dense aggregates of ribosomes (R). The two small holes in the cell represent spaces presumably occupied previously by polyphosphate granules. (w): multilayered complex cell wall typical of blue-green algae. Prefixation with 4 percent glutaraldehyde in 0.1M phosphate buffer, pH 6.8, followed by osmium fixation (Ryter and Kellenberger, 1958) for four hours at room temperature. Dehydration with ethanol followed by propylene oxide. Embedded in Epon 812. Sections post-stained with 0.5 percent uranyl acetate and lead citrate. × 60,000. (Markers indicate 0.1μ.)

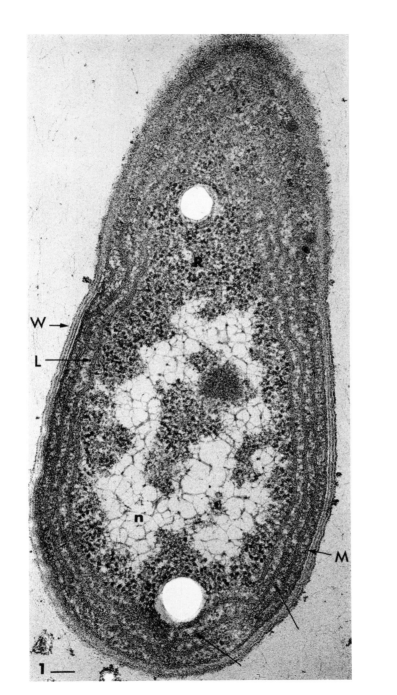

W →

L ——

n

M

1 ——

FIGURES 2 AND 3. *Anacystis* sp.—Thick longitudinal sections. The large granules of phycobiliproteins are very numerous; they are concentrated in the cortical region of the cytoplasm between the cell membrane and the successive lamellae that underlie and parallel it. Note the partly vaporized very electron opaque polyphosphate granules. The cell wall is surrounded by a halo of very fine fibrils.

The cell shown in Figure 3 was extracted with 80 percent acetone after glutaraldehyde prefixation (acetone removes chlorophyll, carotenoids, and presumably most other lipids). As a result, the unit membrane profiles of the lamellae and of the cytoplasmic membrane have disappeared, being replaced by electron transparent spaces. After acetone extraction, the cell suspension has the blue color characteristic of phycocyanin.

FIGURE 2.   Cells prefixed with 3 percent buffered glutaraldehyde followed by osmium fixation. Acetone dehydration and Maraglas embedding. Sections post-stained with lead hydroxide. × 70,000.

FIGURE 3.   Cells fixed with osmium after glutaraldehyde prefixation and acetone extraction. Embedded in Araldite-Epon. Sections post-stained with lead hydroxide. × 80,000.

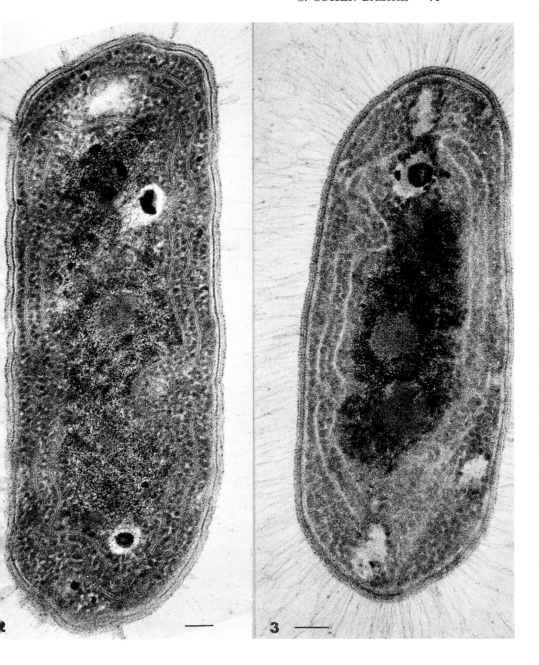

FIGURE 4. Longitudinal section through a large coccoid unicellular blue-green alga in the process of cell division. The lamellae are numerous and mostly cortical; however, some lamellae extend towards the central region of the cell which contains the nuclear material and ribosomes. In favorable parts of the section, the phycobiliprotein granules (g) are seen disposed alternatively on the surface of the lamellae. Many profiles of "polyhedral bodies (pb)" of medium electron density are present in groups in the nucleoplasm.

Note the very complex cell wall with its delicately fibrillar outer layers. The septum is formed at first by annular invagination of the cytoplasmic membrane and a thick deposition of material from the innermost layer of the cell wall, the peptidoglygan layer (M. M. Allen, 1968). Prefixation with 4 percent buffered glutaraldehyde followed by osmium fixation for four hours at room temperature. Acetone dehydration and Maraglas embedding. Sections post-stained with 0.5 percent uranyl acetate and lead citrate. × 60,000.

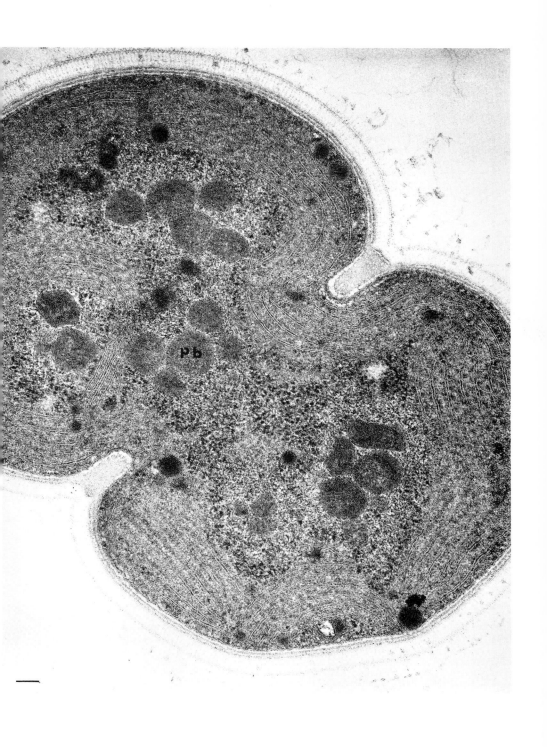

Purple bacteria

Purple bacteria contain one of two chlorophylls known as bacteriochlorophyll *a* and *b*. These pigments have a ring system more reduced than the plant chlorophylls, and consequently they display a major absorption band nearer the infra-red than the red absorption band of the plant chlorophylls (see Table 1). In acetone solution, the major absorption maxima lie at 770 to 772 nm for bacteriochlorophyll *a* and at 795 nm for bacteriochlorophyll *b* (Jensen et al., 1964). *In vivo*, the corresponding bands are displaced into the infra-red region of the spectrum; for bacteriochlorophyll *a*, the absorption maximum lies between 800 and 900 nm, while for bacteriochlorophyll *b* it lies between 1,010 and 1,020 nm. The carotenoids of purple bacteria vary considerably in structure from species to species. They are for the most part aliphatic compounds, often bearing methoxyl groups. Methoxylated carotenoids seem restricted to the purple bacteria (Liaaen Jensen, 1966).

Although many of the purple bacteria are strict anaerobes and obligate phototrophs, some species can grow aerobically in the dark, using the oxidation of organic substrates as a means of obtaining energy. Under such conditions, the photopigment content of the cells becomes very low, since O_2 specifically represses photopigment synthesis (Cohen-Bazire et al., 1957; Lascelles, 1959).

In all purple bacteria, the photosynthetic pigments are associated with more or less deeply infolded unit membranes which can be shown to originate from the cytoplasmic membrane, and which probably at all times remain continuous with it. The form assumed by these infolded membranes varies considerably, but it is characteristic of each species. Broadly speaking, three different types of membranous structures can be recognized. The most common is the vesicular type, which occurs in a majority of the species so far examined. In such purple bacteria, the sites of invagination of the cytoplasmic membrane can be very numerous, and the invaginations take the form of small vesicles about 400 to 500 Å in width, or elongated tubules of the same width, sharply constricted at intervals (Cohen-Bazire and Kunisawa, 1963; Holt and Marr, 1965). When cells rich in photosynthetic pigments are sectioned, the numerous cuts through this complex membrane system give the appearance of a cytoplasm largely filled with electron-transparent circular areas bounded by unit membranes (Figure 5). This type of internal membrane organization has been observed in *Rhodospirillum rubrum, Rhodopseudomonas spheroides,* and *R. capsulatus* among the nonsulfur purple bacteria and in most species of sulfur purple bacteria, including representatives of the genera *Chromatium, Thiocystis, Thiocapsa,* and *Thiospirillum* (Cohen-Bazire, 1963).

In a recently isolated sulfur bacterium, *Thiocapsa pfennigii* (*Thiococcus nov.* sp.) (Eimhjellen et al., 1967) the internal membrane system takes the form of long-branched tubules with a regular diameter of 450 Å, which tend to be arrayed in roughly parallel bundles (Figure 7). A very marked substructure consisting of knobs regularly spaced at a distance of 125 Å from one another is evident in negatively stained preparations (Figure 8). This feature seems to be characteristic of the substructure of the membrane system of purple bacteria that contain bacteriochlorophyll *b;* they include *T. pfennigii* and *Rhodopseudomonas viridis*, which will be described below.

Lastly, certain purple bacteria contain internal membranous intrusions forming multilayered lamellae. In most *Rhodospirillum* species except *R. rubrum* (Cohen-Bazire and Sistrom, 1966; Giesbrecht and Drews, 1962) and in *Ectothiorhodospira* (Trüper, 1968; Remsen et al., 1968; Raymond and Sistrom, 1967), these lamellae occur in disc-shaped stacks, superficially similar to the grana of chloroplasts of higher plants. The stacks of lamellae are formed by the superposition of large flattened vesicles derived from the cytoplasmic membrane and continuous with it. In *Rhodomicrobium* (Boatman and Douglas, 1961; Conti and Hirsch, 1965) and in three *Rhodopseudomonas* species—*R. palustris* (Cohen-Bazire and Sistrom, 1966; Tauschel and Drews, 1967), *R. viridis* (Giesbrecht and Drews, 1966), *R. acidophila* sp. *n.* (Pfenning, 1969)—extended systems of lamellae run parallel and adjacent to the cytoplasmic membrane along the long axis of the cell (Figure 6). In these organisms, there appear to be only one or two specific sites of origin of lamellar intrusions from the cell membrane. Cell division characteristically occurs by budding (Wittenbury and McLee, 1967), and each mother cell appears to retain the preexisting lamellar system, new lamellae being formed during the growth of the daughter cell.

The functional significance of these variations in the disposition of the internal membrane system of purple bacteria is not known. The variations do not appear to be correlated either with specific photopigment patterns or with differences in the mechanisms of photosynthesis. They may conceivably reflect differences in the molecular organization of other membrane compondnts.

The green bacteria

Although the gross mechanism of photosynthesis in green bacteria is very similar to that of many purple bacteria (Pfennig, 1967), the members of this group have a completely different photopigment system. The major chlorophyll is bacteriochlorophyll *c* or *d*, according to the nomenclature proposed by Jensen et al. (1964). These names cor-

FIGURE 5. Longitudinal section through a purple bacterium with vesicular photosynthetic membranes: *Rhodopseudomonas spheroides*. Cells grown at low light intensity (<50 ftc.). The cytoplasm contains numerous isodiametric electron transparent membrane-bounded vesicles (v). The fibrillar nucleoplasm (n) is apparent in various areas of the section between the vesicles. Osmium fixation (R. K.) for two hours. Acetone dehydration. Embedded in Vestopal. Post-stained with lead hydroxide. × 90,000.

FIGURE 6. Longitudinal section through *Rhodopseudomonas palustris* a purple bacterium in which the photosynthetic membranes form an extensive system of tightly packed lamellae. Bacteria of this type characteristically grow and divide by budding, and the cells are typically slipper shaped. The mother cell keeps the entire system of lamellae, a new system being synthesized by the daughter cell. Osmium fixation (R. K.). Acetone dehydration. Embedded in Epon 812. Sections post-stained with lead hydroxide. × 100,000.

FIGURES 7 AND 8. These two photographs of *Thiocapsa pfennigii* have been kindly provided by Dr. K. Eimhjellen, Department of Biochemistry, Technical University of Norway, Trondheim, Norway.

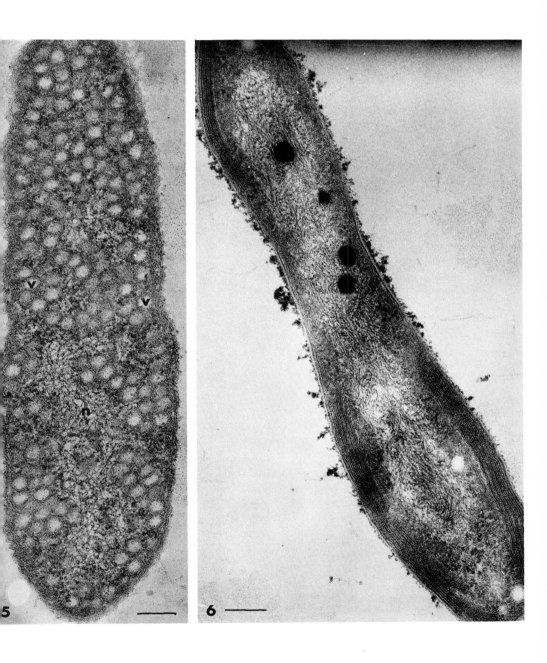

FIGURE 7.   Section through a dividing cell. The tubular photosynthetic membranes characteristic of this species are oriented in large parallel bundles. Each tube is bounded by a unit-type membrane. A large granule of β-hydroxybutyrate polymer (β) occupies the center of one of the future daughter cells and is surrounded by the fine fibrils of nuclear material. Glutaraldehyde prefixation followed by osmium fixation. Vestopal W embedding (Eimhjellen et al., 1967). $\times$ 60,000.

FIGURE 8.   Negatively stained preparation of an osmotically shocked cell of *T. pfennigii* showing the structural continuity of the tubular membranes. The tubes as well as larger pieces of membrane are covered with a regular array of large particles, 125 Å apart. Preparation negatively stained with 1 percent K-phosphotungstate, pH 7. $\times$ 70,000.

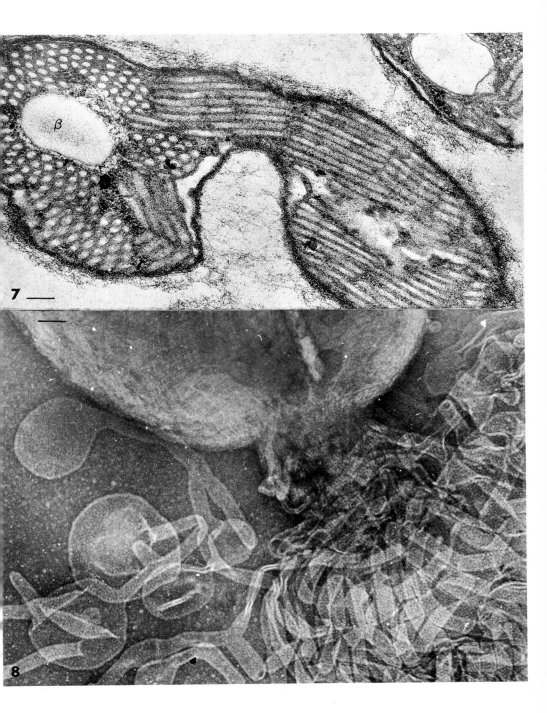

respond to the chlorobium chlorophylls 660 and 650, respectively, of Stanier and Smith (1960). Bacteriochlorophylls *c* and *d* have a ring system on the same oxidation level as that of the plant chlorophylls (Holt, 1966) (see Table 1) and their *in vitro* spectra are accordingly very similar to that of chlorophyll *a*. The marked displacements of the major chlorophyll absorption bands to longer wave lengths *in vivo*, characteristic of purple bacteria, also occur in the green bacteria. The principal *in vivo* chlorophyll absorption bands accordingly lie between 730 and 750 nm, displaced towards a longer wave length by more than 50 nm from the major *in vivo* red absorption bands of the plant chlorophylls.

In addition to either bacteriochlorophyll *c* or *d*, green bacteria always contain bacteriochlorophyll *a* as a minor component of the photopigment system (Olson and Romano, 1962; Jensen et al., 1964). It constitutes some 5 to 10 percent of the total chlorophyll in the cells, and is associated with the photochemical reaction center (Sybesma and Wredenberg, 1966). The green bacteria are therefore unique among photosynthetic organisms by virtue of the fact that there is a molecular difference between their light-gathering chlorophylls and their reaction center chlorophyll. The carotenoids of green bacteria consist of arylcarotenoids, which have an aromatic ring at one or both ends of the linear carbon skeleton (Liaaen Jensen, 1966).

The green bacteria are the only group of photosynthetic organisms, procaryotic or eucaryotic, in which the photopigment system is *not* associated with unit membranes. It is instead housed in specialized organelles known as *chlorobium vesicles* (Cohen-Bazire et al., 1964). These are cigar-shaped structures, about 500 Å wide and 1,200 to 1,500 Å long, which form a cortical layer immediately underlying the cell membrane but distinct from it. Each vesicle is enclosed by a non-unit membrane (single electron-opaque layer about 20-30 Å thick) and appears internally uniform when examined in sections (Figures 9 and 11). Negatively stained preparations of isolated chlorobium vesicles show that they are filled with a reticulum of fibrillar elements (Figure 10).

The localization of the photopigments of green bacteria in these vesicles is demonstrated by the fact that purified vesicles contain most of the cellular chlorophyll (Loeb-Cruden, 1968).

The cells of green bacteria also contain conspicuous internal unit membrane systems which represent invaginations of the cell membrane (Figure 11), and characteristically occur at the site of cell division (Cohen-Bazire et al., 1964). They appear to correspond to the mesosomes which have been described in many nonphotosynthetic bacteria, and are presumed to play a role in nuclear division and transverse wall

formation (FitzJames, 1965; Cohen-Bazire et al., 1966). Partly purified preparations of cytoplasmic membrane and mesosomal fragments
have a negligible chlorophyll content (Loeb-Cruden, 1968).

Gross composition of the procaryotic photosynthetic apparatus

By appropriate methods of physical fractionation, it is possible to
isolate reasonably pure preparations of bacterial membranes and of
chlorobium vesicles. The gross compositions of such preparations are
compared with the composition of chloroplast lamellae (Park, 1966) in
Table 2. The gross composition of the photosynthetic membranes and
the pigment-free membranes of the purple bacterium, *R. spheroides,*
is closely similar to that of chlorophyll-free membranes of the same
species, prepared from cells grown in the presence of oxygen (Gorcheim et al., 1968), and to that of a membrane preparation from the
nonphotosynthetic gram positive bacterium, *Listeria monocytogenes*
(Ghosh and Carrol, 1968; see also Salton, 1967). The lipid content
of all these preparations is about half that of chloroplast lamellae and
of the membranes and vesicles from green bacteria (Loeb-Cruden,
1968). The gross lipid compositions of these preparations are shown
in Table 3. The membrane lipids of *R. spheroides* (Gorcheim, 1968)
and of *Listeria* consist almost entirely of phospholipids, and these
compounds also represent the preponderant lipids of the membrane
fraction of green bacteria (Loeb-Cruden, 1968). The complete absence
of glycolipids from the membranes of purple bacteria should also be

Table 2. Gross Composition of Membranes Fractions
(% dry weight)

| | Chloroplasts (*spinach*) | Purple bacteria (*R. spheroides*) | | Gram positive bacteria (*Listeria monocytogenes*) | Green bacteria (*Chlorobium*) | |
		Pigmented membranes	Aerobic fragments (cytoplasmic membrane)		Membrane fragments	Vesicles
Protein	51	61-63	55	61	29	33
Lipids total[1]	48	27	23	20-22	55	43
"Chlorophyll"	10-11	6-9	0	0	2.3	12-20
Carbohydrate	....	4	16	2	9	16

[1] Including chlorophyll.

FIGURE 9. Thin sections through the green bacterium, *Chlorobium thiosulfatophilum,* a cross section and a longitudinal section through a dividing cell. In this particular strain, the cell wall is covered with numerous pili (these pili are not a general feature of green bacteria). Cell division typically occurs by formation of a septum formed by invagination of the cytoplasmic membrane and centripetal growth of the inner layer of the cell wall, following a process very similar to that observed in blue-green algae. Immediately adjacent to the cytoplasmic membrane (m), a cortical layer of cigar-shaped electron transparent vesicles is visible; these chlorobium vesicles (cv) are bounded by a very thin non-unit membrane 20 Å to 30 Å thick. The central portions of the cells contain the fibrillar nuclear material surrounded by numerous ribosomes. Osmium fixation (R. K.) for two hours. Acetone dehydration. Embedded in methacrylate. Sections post-stained with lead hydroxide. × 80,000.

FIGURE 10. Negatively stained preparation of chlorobium vesicles (cv) partly purified from extracts of *C. thiosulfatophilum.* The large oblong vesicles are filled with a reticulum of fibrillar material; this is most apparent in partly damaged vesicles (arrows). Negatively stained with 1 percent K-phosphotungstate, pH 7. × 120,000.

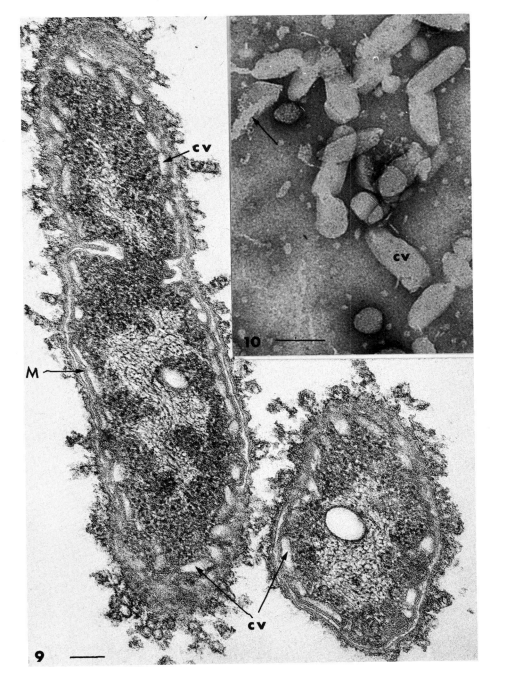

Table 3. Gross Lipid Composition of Membranes Fragments
(% total lipid)

| | Spinach chloroplasts | Green bacteria | | Purple bacteria | Gram positive bacteria |
		Vesicles	Membranes	Pigmented membranes	(*Listeria*)
Phospholipids.....	9	8	80	62	80-85
Glycolipids...........	42	28	17	0	0
	(*galacto-digalacto-*)	(*galacto-*)	(*galacto-rhamno-unknown*)		
"Chlorophyll".....	22	30-50	4	22-30	0

noted. The chlorobium vesicles on the other hand have a completely different lipid composition (Loeb-Cruden, 1968) which closely resembles that of chloroplast lamellae; the phospholipid content is low, and (apart from chlorophyll) the predominant lipid components in these preparations are *galactolipids*. Data on the lipids of membrane fractions from blue-green algae (Levin et al., 1964) indicate that these membranes resemble chloroplasts and chlorobium vesicles; the phospholipid content is low, and the galactolipid content is very high.

In terms of lipid composition, accordingly, the photosynthetic apparatus of the purple bacteria stands apart from all other photosynthetic groups; glycolipids are absent, and the predominant lipids are phospholipids as in the membranes of nonphotosynthetic bacteria (Salton and Freer, 1965; Hagen et al., 1966; Ghosh and Carrol, 1968). Both the blue-green algae and the green bacteria contain large amounts of galactolipids, despite the very marked differences between them with respect to the structure of the photosynthetic apparatus, the photosynthetic pigments, and the type of photosynthesis.

The gas vacuoles as procaryotic organelles

Apart from chlorobium vesicles (confined to one small group of photosynthetic bacteria), the only membrane-enclosed structures so far discovered in procaryotes are *gas vesicles*, constituents of the organelles known as *gas vacuoles* (Cohen-Bazire et al., 1969). Gas vesicles, like chlorobium vesicles, are bounded by a non-unit membrane 20 to 30 Å thick (Figures 11 and 12); they consist of cylinders with conical ends (Figures 11, 12, and 13) which, in the inflated state, are filled with a gas, probably nitrogen. Negatively stained preparations of isolated gas vesicles (Figure 13) show a regular fine striation with a periodicity of 40 Å oriented perpendicularly to their long axis. The system of inflated gas vesicles may in some cases occupy as much as 40 percent of the volume of the cell (Smith and Peat, 1967). They are often arranged in clusters to form one or a few gas vacuoles in each cell. The gas vacuole serves as an organelle of flotation which enables its possessor to occupy a more or less fixed position in a vertical water gradient (Fogg, 1954; Pfennig, 1967). Gas vacuoles occur sporadically in representatives of all three major procaryotic protosynthetic groups and have also been found in a few nonphotosynthetic bacteria.

Conclusions

With increasing knowledge about the structure of chloroplasts and of the cells of blue-green algae, the origin of chloroplasts through the establishment of an endosymbiosis between a procaryotic organism of the blue-green algal type and a nonphotosynthetic eucaryote has become a more and more attractive evolutionary hypothesis. The cell of a blue-green alga, with its procaryotic nuclear structure and 70S ribosomes among which the photosynthetic lamellae are interspersed, is much more closely analogous to the individual chloroplast than to the whole cell of a eucaryotic alga. Contemporary examples of what might constitute evolutionary stages in the transition from free-living blue-green algae to chloroplasts are the so-called cyanelles found in such eucaryotic organisms as *Glaucocystis* and *Cyanophora* (Lefort, 1965; Bourdu and Lefort, 1967). The cyanelles appear to be endosymbiotic unicellular blue-green algae which have lost their cell walls, and hence presumably the possibility of a free-living existence.

The purple and green bacteria appear to be two completely isolated photosynthetic groups, distant both from the blue-green algae and from one another. In structural terms, the photosynthetic apparatus of green bacteria is more complex and specialized than that of either purple bacteria or blue-green algae since the photopigment system of these organisms is structurally segregated within a special organelle, the chlorobium vesicle. Indeed, the green bacteria can, with some justification, be regarded as the most highly differentiated of procaryotic organisms.

With respect to the anatomy of their photosynthetic apparatus, the purple bacteria resemble blue-green algae more closely than they do green bacteria. However, the internal unit membranes which bear the photopigment system of purple bacteria is similar in chemical composition to the membranes of nonphotosynthetic bacteria, as shown most strikingly by the complete absence of galactolipids.

FIGURE 11.   Section through a dividing cell of *Pelodictyon clathratiforme,* a species of green bacteria which contains gas vacuoles. The chlorobium vesicles (cv) are a medium electron density (when the cells are embedded in epoxy resins) and are surrounded by a thin non-unit membrane more clearly visible than in Figure 9. Note the very large mesosome (m) at the site of cell division. A gas vesicle (gv) is sectioned longitudinally in the upper part of the cell, showing the characteristic shape of gas vesicles (a cylinder with two conical ends) and the very thin non-unit membrane that encloses them. Osmium fixation (R. K.) for two hours. Acetone dehydration. Embedded in Maraglas. Section post-stained with lead hydroxide. × 80,000.

FIGURE 12.   Picture showing part of a cell of *Oscillatoria agardhii,* var. *suspensa,* a blue-green alga that contains gas vacuoles. The gas vacuoles are composed of many gas vesicles grouped in bundles. The section shows the gas vesicles cut at various angles. Their shape and structure is identical with the gas vesicle shown in FIGURE 11 in a green bacterium. Osmium fixation for four hours. Embedded in Maraglas. Post-stained with lead hydroxide. × 60,000.

FIGURE 13.   Negatively stained preparation of gas vesicles isolated from *Oscillatoria agardhii.* Note their general shape and the fine striations covering their surface. Negative stain: 1 percent Uranyl acetate. × 150,000.

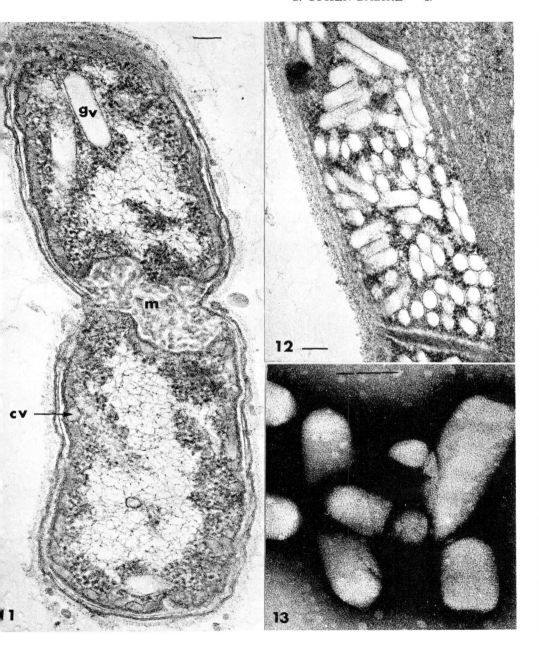

Literature Cited

Allen, M. M. 1968. Ultrastructure of the cell wall and cell division of unicellular blue-green algae. J. Bact., *96*:842-852.

Boatman, E. S., and H. C. Douglas. 1961. Fine structure of the photosynthetic bacterium, *Rhodomicrobium vannielii.* J. Biophys. Biochem. Cytol., *11*:469-483.

Bourdu, M., and M. Lefort. 1967. Structure fine, observée en cryodécapage, des lamelles photosynthétiques des cyanophycées endosymbiotiques: *Glaucocistis nostochinearum* Itzigs et *Cyanophora paradoxa* Korschikoff. C. R. Acad. Sc. Paris, *265*:37-40.

Cohen-Bazire, G. 1963. Some observations on the organization of the photosynthetic apparatus in purple and green bacteria. In *Bacterial Photosynthesis.* H. Gest, A. San Pietro, and L. P. Vernon, eds. Yellow Springs, Ohio: The Antioch Press, pp. 89-110.

Cohen-Bazire, G., and R. Kunisawa. 1963. The fine structure of *Rhodospirillum rubrum.* J. Cell Biol., *16*:401-419.

Cohen-Bazire, G., and M. Lefort. 1968. Structure granulaire du chromotoplasma des cyanophycées. Fourth European Regional Conference on Electron Microscopy, Rome, pp. 395-396.

Cohen-Bazire, G., R. Kunisawa, and J. S. Poindexter. 1966. The internal membranes of *Caulobacter crescentus.* J. Gen. Microbiol., *42*:301-308.

Cohen-Bazire, G., N. Pfennig, and R. Kunisawa. 1964. The fine structure of green bacteria. J. Cell. Biol., *22*:207-225.

Cohen-Bazire, G., N. Pfennig, and R. Kunisawa. 1969. A comparative study of the structure of gas vacuoles. J. Bact., *100*:1049-1061.

Cohen-Bazire, G., and W. R. Sistrom. 1966. The procaryotic photosynthetic apparatus. In *The Chlorophylls,* L. P. Vernon and G. R. Seely, eds. New York: Academic Press, Inc., pp. 313-341.

Cohen-Bazire, G., W. R. Sistrom, and R. Y. Stanier. 1957. Kinetic studies of pigment synthesis by non-sulfur purple bacteria. J. Cell. Comp. Physiol., *49*:25-68.

Conti, S. F., and P. Hirsh. 1965. Biology of budding bacteria. III. Fine structure of *Rhodomicrobium* and *Hyphomicrobium* spp. J. Bact., *89*:503-512.

Edwards, M. R., D. S. Berns, W. C. Ghiorse, and S. C. Holt. 1968. Ultrastructure of the thermophilic blue-green alga. *Synechococcus lividus.* J. Phycol., *11*: 283-298.

Eimhjellen, K. E., H. Steensland, and J. Traetteberg. 1967. A *Thiococcus* sp. nov. gen., its pigments and internal membrane system. Arch. Mikrobiol., *59*:82-92.

FitzJames, P. C. 1965. A consideration of bacterial membranes as the agent of differentiation. In *Function and Structure in Microorganisms.* Symp. Soc. Gen. Microbiol., *15*:369-378.

Fogg, G. E. 1941. The gas vacuoles of the *Myxophyceae* (*Cyanophyceae*). Biol. Rev., *16*:205-217.

Gantt, E., and S. F. Conti. 1966. Phycobiliprotein localization in algae. In *Energy Conversion by the Photosynthetic Apparatus.* Brookhaven Symposia in Biology, *19:*393-405.

Gantt, E., and S. F. Conti. 1969. Ultrastructure of blue-green algae. J. Bact., *97:*1486-1493.

Ghosh, B. K., and K. K. Carrol. 1968. Isolation, composition, and structure of membrane of *Listeria monocytogenes.* J. Bact., *95:*688-699.

Giesbrecht, P. and G. Drews. 1962. Elecktronenmikroskopische Utersuchungen über die Entwicklun der "chromatophoren" von *Rhodospirillum molischianum* Giesberger. Arch. Mikrobiol., *43:*152-161.

Giesbrecht, P., and G. Drews. 1966. Über die Organisation und die makromolekulare Architektur der Thylakoide "lebender" Bacterien. Arch. Mikrobiol., *54:*297-330.

Gorcheim, A. 1968. The separation and identification of the lipids of *Rhodopseudomonas spheroides.* Proc. Roy. Soc. B., *170:*279-297.

Gorcheim, A., A. Neuberger, and G. H. Tait. 1968. The isolation and characterization of subcellular fractions from pigmented and unpigmented cells of *Rhodopseudomonas spheroides.* Proc. Roy. Soc. B., *170:*229-246.

Hagen, P. O., H. Goldfine, and P. J. LeB. Williams. 1966. Phospholipids of bacteria with extensive intracytoplasmic membranes. Science, *151:*1543-1544.

Holt, A. S. 1966. Recently characterized chlorophylls. In *The Chlorophylls.* L. P. Vernon and G. R. Seely, eds. New York: Academic Press, Inc., pp. 111-118.

Holt, S. C., and A. G. Marr. 1965. Location of chlorophyll in *Rhodospirillum rubrum.* J. Bact., *89:*1402-1420.

Jensen, A., O. Aasmundrud, and K. E. Eimhjellen. 1964. Chlorophylls of photosynthetic bacteria. Biochem. Biophys. Acta. *88:*466-479.

Lang, N. J. 1968. The fine structure of blue-green algae. Ann. Rev. Microbiol. *22:*15-46.

Lascelles, J. 1959. Adaptation to form bacteriochlorophyll in *Rhodopseudomonas spheroides:* Changes in activity of enzymes concerned in pyrrole synthesis. Biochem. J., *72:*508-518.

Lefort, M. 1965. Sur le chromatoplasma d'une cyanophycée endosymbiotique. *Glaucostis nostochinearum* Itzigs. C. R. Acad. Sc. Paris, *261:*233-236.

Levin, E., W. J. Lennarz, and K. Bloch. 1964. Occurrence and localization of α-linolenic acid containing galactolipids in the photosynthetic apparatus of *Anabaena variabilis.* Biochim. Biophys. Acta., *84:*471-474.

Liaaen Jensen, S. 1966. Recent studies on the structure and distribution of carotenoids in photosynthetic bacteria. In *Biochemistry of Chloroplasts I.* Goodwin, T. H. ed., London: Academic Press, pp. 437-441.

Loeb-Cruden, D. 1968. Structure and function in photosynthetic procaryotic organisms. Ph.D. thesis, University of California, Berkeley, Calif.

Olson, J. M., and C. A. Romano. 1962. A new chlorophyll from green bacteria. Biochim. Biophys. Acta., *59:*726-728.

Pankratz, H. S., and C. C. Bowen. 1963. Cytology of blue-green algae. I. The cells of symploca muscorum. Am. J. Bot., *50:*387-399.

Park, R. B. 1966. Chloroplast structure. In *The Chlorophylls*. L. P. Vernon and G. R. Seely, eds., New York: Academic Press, Inc., pp. 283-311.

Pfennig, N. 1967. Photosynthetic bacteria. Ann. Rev. Microbiol., *21*:285-324.

Pfennig, N. 1969. *Rhodopseudomonas acidophila,* sp. n., a new species of the budding purple nonsulfur bacteria. J. Bact., *99*:597-602.

Raymond, J. C., and W. R. Sistrom. 1967. The isolation and preliminary characterization of a halophilic photosynthetic bacterium. Arch. Mikrobiol., *59:* 255-268.

Remsen, C. C., S. W. Watson, J. B. Waterbury, and H. G. Trüper. 1968. Fine structure of *Ectothiorhodospira mobilis* Pelsh. J. Bact., *95:*2374-2392.

Ryter, A. 1968. Association of the nucleus and the membrane of bacteria: a morphological study. Bact. Rev., *32:*39-54.

Ryter, A., and E. Kellenberger. 1958. Etude au microscope electronique des plasmas contenant de l'acide désoxyribonucléique. Z. Naturforsch., *13b:* 597-605.

Salton, M. R. J. 1967. Structure and composition of bacterial membranes. In *Protides of the Biological Fluids, 15:*279-288. Amsterdam: Elsevier Publishing Co.

Salton, M. R. J., and J. H. Freer. 1965. Composition of the membranes isolated from several gram-positive bacteria. Biochim. Biophys. Acta., *107:*531-538.

Smith, R. V., and A. Peat. 1967. Comparative structure of the gas-vacuoles of blue-green algae. Arch. Mikrobiol., *57:*111-122.

Stanier, R. Y., and J. H. C. Smith. 1960. The chlorophylls of green bacteria. Biochim. Biophys. Acta., *41:*478-484.

Sybesma, C., and W. J. Wredenberg. 1964. Kinetics of light-induced cytochrome oxidation and P840 bleaching in green photosynthetic bacteria under various conditions. Biochim. Biophys. Acta., *88:*205-207.

Tauschel, H. D., and G. Drews. 1967. Thylakoidmorphogenese bei *Rhodopseudomonas palustris*. Arch. Mikorbiol., *59:*381-404.

Trüper, H. G. 1968. *Ectothiorhodospira mobilis* Pelsh, a photosynthetic sulfur bacterium depositing sulfur outside the cells. J. Bact., *95:*1910-1920.

Wittenbury, R., and A. G. McLee. 1967. *Rhodopseudomonas palustris* and *Rh. viridis*—Photosynthetic budding bacteria. Arch. Mikrobiol., *59:*324-334.

ACKNOWLEDGMENTS: The work reported in this article was supported by a grant from the National Science Foundation to Professor Michael Doudoroff.

The author wishes to acknowledge the valuable help given by Miss Riyo Kunisawa and by the staff of the Electron Microscope Laboratory, particularly Mr. Philip Spencer for the photographic work.

An Approach Towards Defining the Role of Chloroplast DNA in the Reproduction and Differentiation of Chloroplasts in Higher Plants

S. G. WILDMAN
Department of Botanical Sciences, Molecular Biology Institute
University of California, Los Angeles

CHLOROPLASTS found in higher plants such as tobacco and spinach are biphasic organelles. When viewed in living cells by phase microscopy, chloroplasts are seen to be composed of a *stationary component* containing grana, surrounded by a nearly transparent material, the *mobile phase,* which is constantly changing shape and exhibiting various interactions with the mitochondria and cytoplasmic network of the living cell (*11, 28, 29*).[1] Chlorophyll is confined to the stationary component. Figure 1 is a scale model of a spinach leaf mesophyll chloroplast which attempts to depict the biphasic organization of the chloroplast, utilizing information derived from both light and electron microscopy. The essential feature of this model is the probable organization of the thylakoid system in the form of a stack of flattened pancakes (*22*) surrounded and interpenetrated by the mobile phase. Speculations have been presented as to the significance of this kind of organization in relation to compartmentalization of the light transduction system and its separation from the enzyme systems responsible for CO_2 fixation during photosynthesis (*28*).

Methods have been developed for isolating chloroplasts from leaves with their mobile phase intact and for selectively removing the mobile phase from the thylakoid system of lamellae (*5, 10*). While leaves contain both 80S and 70S ribosomes (*17*), analysis of the macromolecular composition of the mobile phase shows it to contain the 70S ribosomes and 18S Fraction I protein (carboxydismutase) together with a mixture of proteins of 6S or smaller, some of which have been identified as participants with Fraction I protein in the Calvin cycle of CO_2 fixation. The 70S chloroplast ribosomes have

---

[1] Italic numbers in parentheses refer to Literature Cited, page 104.

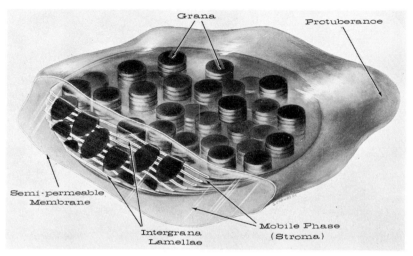

FIGURE 1. Scale model of average chloroplast found in mesophyll cells of mature spinach leaf. Diameter of grana $= 0.5\mu$. From 200 to 600 chloroplasts are present in palisade parenchyma cells with diameters ranging from 6 to 12μ. Chloroplasts occupy about 50 percent of the total area of protoplasm. (After Francki et al. (5), reproduced with permission of *Biochemistry*.)

physico-chemical and morphological properties closely similar to ribosomes obtained from microorganisms such as *E. coli* and blue-green algae (1). Leaves have also been demonstrated to contain five species of ribosomal RNA (16, 23): 16S and 23S RNA associated with the chloroplast ribosomes and 18S and 26S RNA associated with the 80S cytoplasmic ribosomes together with 5S RNA, perhaps associated with both species of ribosomes.

Isolated chloroplasts with intact mobile phase incorporate amino acids into peptides, utilizing polysomes composed of 70S ribosomes (2). About one half of the chloroplast ribosomes are readily extracted with the mobile phase; the remainder are so tightly associated with the thylakoid system that detergents are required for their release and then only about 25 percent can be released without concomitant release of chlorophyll resulting from disorganization of the thylakoids (3).

Chloroplast DNA

Studies of a great variety of photosynthetic organisms including several higher plants have shown chloroplasts to contain a unique DNA. In tobacco leaves, chloroplast DNA is tightly associated with the thylakoid system, constitutes about 9 percent of the total DNA in the leaves, and amounts to about 5 x 10^{-15} g of DNA per chloroplast (24). Associated with the chloroplast DNA and the thylakoid systems are a DNA polymerase with the potential for replication of chloroplast DNA (21, 26) and an RNA polymerase with the potential of transcribing chloroplast DNA into m-RNA (12, 20, 25). Thus, with 70S ribosomes, chloroplasts appear to have all of the necessary enzyme systems which would be required for replication, transcription, and translation of the chloroplast DNA, the latter being estimated to contain enough information to code for more than a thousand different proteins containing as many as two hundred amino acids in their primary structures (13). Properties which serve to distinguish tobacco chloroplast DNA from nuclear and mitochondrial DNA are listed in Table 1.

Table 1. Summary of Properties Which Distinguish Between Nuclear, Chloroplast, and Mitochondrial DNA's Obtained from Tobacco Leaves

Property	Chloroplast	Nuclear	Mitochondria
Percent of total leaf DNA	9	90 +	> 1
DNA per organelle	9×10^{-15} g	100×10^{-15} g	—
Density in CsCl	1.700-1.702	1.696-1.698	1.724
Length	30-150μ	Unknown	2-12 μ
Renaturation	Complete	Incomplete	Complete
5-methyl cytosine	Absent	2-5%	—
Histones	Absent	Present	—
Transcription: $\mu\mu$ moles RNA/hr./μg DNA	40	1.2	—

The question remains as to what role chloroplast DNA performs in the development and reproduction of chloroplasts, a question being asked in several laboratories throughout the world. In the case of our own work with tobacco, some evidence is at hand which can be interpreted as suggesting that chloroplast DNA does function during the life of tobacco chloroplasts.

Regulation of chloroplast RNA and protein synthesis by tobacco mosaic virus

Conditions have been found whereby tobacco mosaic virus (TMV) will accumulate in small tobacco leaves at exponential rates. The amount of virus rises to levels of 5 mg TMV/gram tissue within 72 hours after inoculation (8). Recent work indicates that TMV interferes with macromolecular metabolism of chloroplasts (9). About

60 hours after TMV infection, leaves were supplied with ^{32}P and a mixture of ^{3}H-amino acids, and extracts suitable for resolution and analysis of macromolecular components were made about 8 hours later. Analysis revealed that ^{32}P became incorporated into TMV-RNA and the 18S and 26S ribosomal RNA's derived from the 80S cytoplasmic ribosomes, but not into the 16S and 23S RNA's contained in the chloroplast ribosomes. In addition, TMV protein and cytoplasmic proteins incorporated the ^{3}H-amino acids, but there was almost negligible incorporation of the ^{3}H-amino acids into chloroplast proteins such as Fraction I protein. Extensive incorporation of ^{32}P into ribosomal RNA's from both 70S and 80S ribosomes as well as ^{3}H-amino acids into chloroplast and cytoplasmic proteins occurred in uninfected leaves. Further comparative experiments showed isolated chloroplasts from TMV-infected leaves to have a greatly reduced capacity for amino acid incorporation into peptides and nucleoside triphosphate incorporation into m-RNA, the latter utilizing chloroplast DNA as template. Of more significance was the finding that the reduced amount of m-RNA synthesized by isolated chloroplasts from TMV infected leaves had different AMP/GMP and UMP/GMP ratios than m-RNA synthesized by isolated chloroplasts from healthy leaves. Thus, the argument could be advanced that both ribosomal and m-RNA production from DNA in chloroplasts was switched off by an, as yet, undefined regulator arising during the grand period of TMV accumulation elsewhere in the cell. Regulation resulted in the cessation of r-RNA synthesis together with a reduced amount and different kinds of m-RNA's being formed. Reduction in m-RNA was the probable cause of reduction in protein synthesizing capacity both *in vivo* and *in vitro*. This evidence is a favorable indication that chloroplast DNA does function in the formation of chloroplast RNA's and proteins in healthy leaves.

Hybridization of ribosomal RNA with chloroplast DNA

In furtherance of the ultimate objective of finding out whether base sequences present in chloroplast DNA are reflected in the arrangement of amino acids in the primary structures of chloroplast proteins, evidence is being sought as to whether chloroplasts contain specific t-RNA's different from those in the cytoplasm, whether isolated chloroplasts can synthesize peptides related to known chloroplast proteins such as Fraction I protein, and whether chloroplast DNA contains base sequences which can hybridize with ribosomal RNA from chloroplast ribosomes. In the case of ribosomal RNA, chloroplast DNA hybridizes with ribosomal RNA from 70S chloroplast ribosomes (0.5 percent) (*19, 27*) but not with RNA from 80S cytoplasmic ribosomes; the latter hybridizes with nuclear DNA (0.3 percent). Surprisingly, chloroplast ribosomal RNA also hybridizes with nuclear DNA (0.1 percent) (*27*). In fact, the informational content of nuclear DNA related to chloroplast ribosomal RNA is about three times greater than that contained in chloroplast DNA since the latter constitutes only 9 percent of the total cellular DNA. On a per nucleus per chloroplast basis, the information in a nucleus capable of coding for chloroplast RNA is about a thousand times greater than that contained in the DNA of a chloroplast. There appear to be about 800 cistrons in nuclear DNA complementary to chloroplast ribosomal RNA, compared to only about 8 cistrons in chloroplast DNA. Apparently, only about 1 percent of the chloroplast DNA contains nucleotide sequences complementary to chloroplast ribosomal RNA, suggesting that much of the potential information in chloroplast DNA is available to code for other purposes. In any event, chloroplast DNA does appear to have the potential and specificity to function in the synthesis of the ribosomal RNA of chloroplast ribosomes.

Is the information in nuclei for chloroplast ribosomal RNA utilized during the development of chloroplasts or is it a nonfunctional relic of past evolution? There is evidence that nuclear and chloroplast DNA's share common nucleotide sequences as detected by hybridization (*18, 26*). Perhaps the cistrons in nuclear DNA complementary to chloroplast ribosomal RNA arose in the manner depicted in Figure 2. The thylakoid system for photosynthesis first appeared in a unicellular blue-green algae which did not possess organelles in its structure other than a thylakoid containing chlorophyll. DNA was organized as in present-day microorganisms, but a portion of the DNA contained a sequence of nucleotides which provided the information necessary for synthesis of the thylakoid, and this sequence of nucleotides constituted primeval chloroplast DNA. Ensuing evolution to the level of complexity of a green alga such as *Chlorella* resulted in the encapsulation of nuclear DNA within a nucleus and the thylakoid system within the organelle we recognize as a chloroplast. At some stage in the evolution of thylakoids into chloroplasts, the DNA necessary for thylakoid synthesis became detached but still retained the ability to replicate and is now the source of information for the formation of chloroplasts independent of control by nuclear DNA. The portion of DNA which gave rise to autonomous chloroplast DNA still remains in nuclear DNA as a redundant, nonfunctional relic of evolution. However, it seems certain that nuclear DNA does have a role in the development of chloroplasts as attested by the many Mendelian mutants which affect chlorophyll biosynthesis and other aspects of

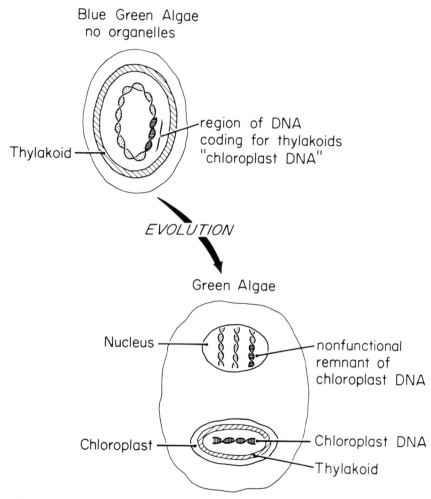

Blue Green Algae
no organelles

region of DNA
coding for thylakoids
"chloroplast DNA"

Thylakoid

EVOLUTION

Green Algae

Nucleus

nonfunctional
remnant of
chloroplast DNA

Chloroplast

Chloroplast DNA

Thylakoid

FIGURE 2.   Possible origin of autonomous chloroplast DNA during evolution leading to organelles within cells.

chloroplasts (*14*). In addition, some work on a cytoplasmic mutant of tobacco (*15*) persuades me that a more attractive hypothesis may be one that invokes a function for nuclear DNA as containing information required for the development of the mobile phase of chloroplasts, whereas information in chloroplast DNA is utilized for the formation of the thylakoids.

Organization and macromolecular composition of defective chloroplasts found in a cytoplasmic mutant of tobacco

A mutant tobacco plant has been obtained which can be propagated by seeds and where a random, bizarre variegation of the leaves occurs. The variegation character is transmitted only through the maternal line. Microscopic examination of living mesophyll cells from variegated areas revealed mixed populations of normal and defective chloroplasts in the same cells. The degree of variegation is correlated with the ratio of normal to defective chloroplasts, the latter being devoid of chlorophyll. When defective chloroplasts predominate, the tissue is almost white in appearance. No area of the leaves, however green, is entirely devoid of defective chloroplasts intermingled with normal chloroplasts. The defective chloroplasts have never been detected in wild type tobacco. As observed by phase microscopy, defective chloroplasts appear to be equivalent to the mobile phase, showing the same changes in shape, having about the same extent, and interacting with mitochondria as is the case with the mobile phase of normal chloroplasts. Furthermore, starch grains appear in the defective chloroplasts, implying that enzymes are present for synthesis of starch and that some enzymes of the Calvin cycle are present.

Isolation and analysis of defective chloroplasts revealed the presence of chloroplast DNA in amounts equivalent to those of normal chloroplasts as well as evidence for a DNA polymerase. More striking was the finding of the same amount of 70S chloroplast ribosomes as are present in normal chloroplasts from wild type leaves. Fraction I protein was also present in defective chloroplasts and displayed the same specific carboxydismutase activity as the enzyme from wild type chloroplasts. However, the amount of Fraction I protein was reduced to about 25 percent of that found in normal chloroplasts. The presence of starch as well as carboxydismutase suggests that most or all of the enzymes of the Calvin cycle may be present in defective chloroplasts. RNA polymerase in defective chloroplasts was almost inactive, although the RNA polymerase in the nuclei from the same variegated tissue displayed the same level of activity as found in nuclei from wild type plants. Thus, a plausible interpretation of these findings would be to assume that the chloroplast DNA has been impaired in function to the extent that a thylakoid system fails to develop, or only very poorly, within a mobile phase. A mechanism for repression of the DNA can be ruled out because defective chloroplasts exist in the same cells as normal chloroplasts and it would be hard to imagine how a repressor could shut off transcription in some chloroplasts and not in others. The more probable explanation is that chloroplast DNA has suffered a mutation which prevents transcription of the entire genome, and the mechanics of chloroplast reproduction during cellular division permit the mutated DNA to be replicated and transmitted from cell to cell together with wild type chloroplast DNA. On the basis of a recent report on electron microscope observations of dividing tobacco meso-phyll cells, it is possible to imagine how mutated chloroplast DNA could be transmitted together with wild type chloroplast DNA to pro-duce cytoplasmic mutants displaying variegated leaves resulting from defective chloroplasts.

Scheme for chloroplast reproduction in tobacco plants

Knowledge of the precise nature of the stages in chloroplast re-production and development in higher plants is extremely scanty. Ap-parently, well-developed chloroplasts have not been identified in egg cells of higher plants or in seeds, and, therefore, independent, chloro-phyll-containing organelles do not appear to be transmitted from generation to generation. However, chloroplast DNA has been identi-fied in seeds (7), and it may be suspected that chloroplast DNA is the agent which is transmitted and contains information necessary for the complete formation of each new generation of chloroplasts upon germination of seeds, emergence of apical meristems, and the con-tinuous formation of new leaf primordia from the meristem as the plant grows. Intensive observations by electron microscopy of leaf mesophyll cells in the process of being formed within leaf primordia should provide important clues as to the mechanics of chloroplast reproduction. Recently, some observations of dividing tobacco leaf mesophyll cells have been published (4), and from these, it is possible to arrive at a plausible notion of how chloroplasts may reproduce and differentiate.

Figure 3a is a section of two tobacco leaf mesophyll cells about to complete division into four cells. Mitosis has occurred, and only the cell plate needs to be completed in order to finish the cell division process. What strikes me as significant is that the chloroplasts (and the mitochondria) are already well-differentiated organelles. The chloroplasts have conspicious thylakoid systems, including grana stacks, and the size of the organelles is already one third to one half of the size they will attain in the fully mature leaf. In other observa-tions embracing all stages in mitosis from early prophase through metaphase up to the late stage shown in Figure 3, it does not appear that the chloroplasts are different in form from their appearance in this section. Yet, during some of these stages, the number of chloroplasts per cell must have increased because otherwise some mesophyll cells would appear without chloroplasts, which is not the case in normal tobacco leaves. Thus, in these dividing cells, it is not obvious that structures such as proplastids without chlorophyll or prolamellar bodies containing protochlorophyllide found in completely etiolated leaves, such as those shown in Figure 3b, appear as intermediate stages during the formation of new chloroplasts. In addition, I was unable to find figures which would persuade me that chloroplasts reproduce by fission in the sense that the thylakoid systems became equally proportioned into two daughter chloroplasts.

While the entire three-dimensional reconstruction of sectioned chloroplasts would be required for certainty, it appeared to me that protuberances project from the chloroplasts, and that protuberances have an appearance that might be expected for the mobile phase found in mature chloroplasts. Occasionally, I thought I could see a single thylakoid within a protuberance well-separated from the main thylakoid system in the body of the chloroplast. It is this condition which leads me to propose the mechanism for chloroplast reproduction depicted in Figure 4.

On the basis of work on the cytoplasmic mutant, the presumption is made that information contained in nuclear DNA controls the synthesis of the macromolecular constituents of the mobile phase of chloroplasts. Work with normal chloroplasts shows that ribosomes exist both in the mobile phase and tightly bound to the thylakoids, and it may be presumed that the latter are involved in synthesis of the lipoproteins of the thylakoids and that they are programmed with m-RNA transcribed from chloroplast DNA. Thus, we can envisage that as cells in apical meristems divide and differentiate to become the origin of leaf primordia, structures akin to defective chloroplasts consisting of mobile phase containing chloroplast DNA are formed as the consequence of nuclear DNA control. Activation of chloroplast DNA transcription results in the formation of thylakoids within the mobile phase, and this act is independent of control by nuclear DNA.

(Legends for figures on pages 100 and 101)

FIGURE 3. Electron micrographic visualization of tobacco mesophyll cells in the process of division (A) and prolamellar bodies found in etiolated sugar cane leaves (B). (A, from Cronshaw and Esau (4), reproduced with permission of *Protoplasma;* B, courtesy of Dr. W. M. Laetsch, University of California, Berkeley.)

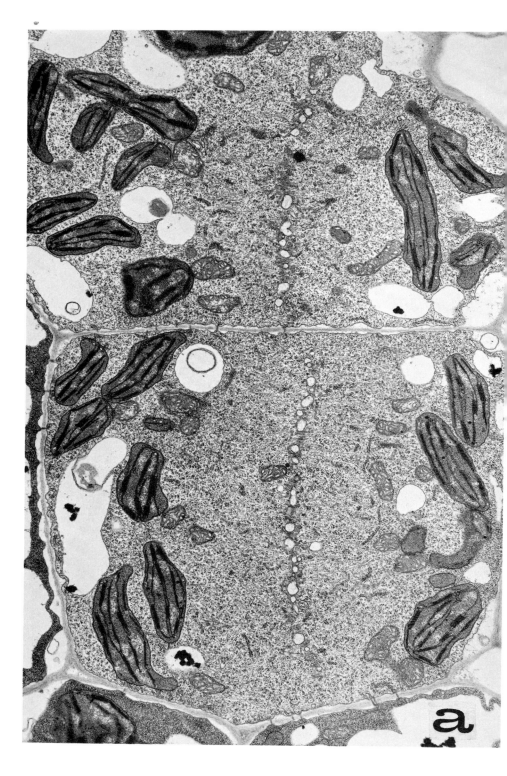

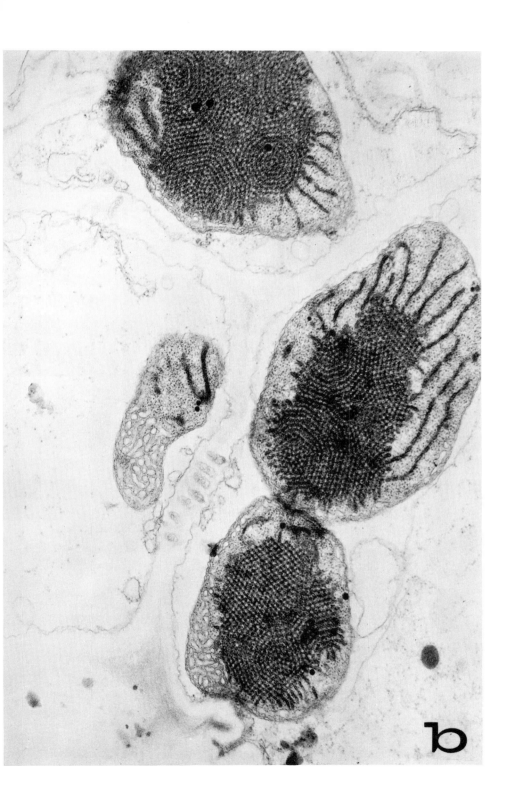

b

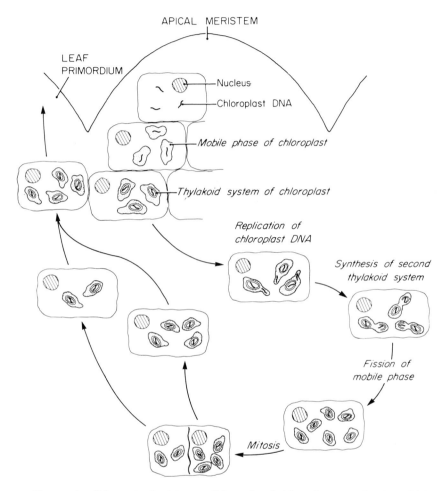

FIGURE 4. Scheme to depict a possible mode of chloroplast reproduction within apical meristems of higher plants. Apparent replication of chloroplast DNA during leaf development as detected by isotope incorporation has been reported (6) and confirmed by our unpublished experiments.

After formation of the thylakoid system, replication of chloroplast DNA occurs, and the newly formed chloroplast DNA can migrate into mobile phase protuberances and reinstitute the synthesis of a second thylakoid system. After formation of the second thylakoid system, a region of mobile phase between the two thylakoid systems separates to yield two separate chloroplasts.

Separation of the mobile phase surrounding two stationary components, or fusion of two chloroplasts via interconnecting mobile phases, is readily observed in living mesophyll cells containing mature chloroplasts. Following chloroplast reproduction, mitosis of the nucleus occurs. Before the new cell wall forms to separate the two nuclei into separate cells, the newly reproduced chloroplasts can distribute unequally into the forthcoming new cells, a necessary condition because it is known that the number of chloroplasts per mature mesophyll cell is variable. With the formation of the new cells, further cell divisions preceded by chloroplast reproduction occur until a point is reached where nuclear division ceases and the cells elongate into the form we recognize as the mesophyll of mature leaves. Even during the process of elongation, it would seem likely that some new chloroplasts could appear by the mechanism proposed to yield the ultimate complement of 200 to 600 chloroplasts per palisade parenchyma cell found in mature leaves.

Explanation for variegated leaves inherited in a non-Mendelian manner is found by merely substituting mutated chloroplast DNA in the scheme. The mutated DNA together with wild type DNA is transmitted through seeds, and both DNA's appear in the mobile phase structures in the apical meristems. The mutated chloroplast DNA does not possess the correct information for synthesis of a normal thylakoid system, but the DNA can still be replicated and transmitted within the mobile phase into new cells. The ratio of mutated DNA to wild type DNA in a dividing mesophyll cell and the statistics of distribution of the two kinds of DNA into the newly formed cells determines the extent of variegation seen on the developing leaves.

Whether or not this suggestion for the role of DNA in chloroplast reproduction will survive in the testing ground of future experiments—or worse, that I have overlooked experimental evidence that would negate the notions at the outset—it is nevertheless clear to me that great opportunities now exist for coming to grips with the problem of what chloroplast DNA is doing.

ACKNOWLEDGMENTS: While disclaiming any responsibility on their part for the speculations presented here, nothing could have been engendered without the experimental facts provided by my numerous talented collaborators. I should like to mention especially Drs. K. K. Tewari, William Burton, Jane Chen, Catherine Liao, Atsushi Hirai, Tasani Hongladarom, Shigeru Honda, N. K. Boardman, Richard Francki, and Donald Spencer. I am also grateful for the financial support provided by the Atomic Energy Commission and the United States Public Health Service.

Literature Cited

1. Boardman, N. K., R. I. B. Francki, and S. G. Wildman. 1966. Protein synthesis by cell-free extracts of tobacco leaves. III. Comparison of the physical properties and protein synthesizing activities of 70S chloroplast and 80S cytoplasmic ribosomes. J. Mol. Biol., $17:470$-489.

2. Chen, Jane L., and S. G. Wildman. 1967. Functional chloroplast polyribosomes from tobacco leaves. Science, $155:1271$-1273.

3. Chen, Jane L., and S. G. Wildman. 1970. "Free" and membrane-bound ribosomes, and nature of products formed by isolated tobacco chloroplasts incubated for protein synthesis. Biochim. Biophys. Acta, $209:207$.

4. Cronshaw, J., and K. Esau. 1967. Cell division in leaves of *Nicotiana.* Protoplasma, $65:1$.

5. Francki, R. I. B., N. K. Boardman, and S. G. Wildman. 1965. Protein synthesis by cell-free extracts from tobacco leaves. I. Activity in relation to chloroplast structure. Biochemistry, $4:865$.

6. Green, B. R., and M. P. Gordon. 1966. Replication of chloroplast DNA of tobacco. Science, $152:1071$.

7. Green, B. R., and M. P. Gordon. 1967. The satellite DNA's of some green plants. Biochim. Biophys. Acta, $145:378$.

8. Hirai, A., and S. G. Wildman. 1967. Intracellular site of assembly of TMV-RNA and protein. Virology, $33:467$-473.

9. Hirai, A., and S. G. Wildman. 1969. Effect of TMV multiplication on RNA and protein synthesis in tobacco chloroplasts. Virology, $38:73$-82.

10. Honda, S. I., T. Hongladarom, and G. Laties. 1966. A new isolation medium for plant organelles. J. Exp. Biol., $17:460$.

11. Hongladarom T., S. Honda, and S. G. Wildman. 1965. Appearance and behavior of organelles in living plant cells. 26 minute, natural color, 16 mm cine-photomicrographic film with commentary on sound track. Extension Media Center, University of California, Berkeley.

12. Kirk, J. T. O. 1964. DNA-dependent RNA synthesis in chloroplast preparations. Biochem. Biophys. Res. Comm., $14:393$.

13. Kirk, J. T. O. 1966. Nature and function of chloroplast DNA. In *Biochemistry of Chloroplasts,* T. W. Goodwin, ed., Vol. 1, page 319. London: Academic Press.

14. Kirk, J. T. O., and R. A. E. Tilney-Bassett. 1967. *The Plastids.* W. H. Freeman & Co. Ltd.

15. Liao, Catherine L. 1968. Macromolecular composition of defective chloroplasts from a cytoplasmic mutant of tobacco. Thesis, University of California, Los Angeles.

16. Loening, V. E., and J. Ingle. 1961. Diversity of RNA components in green plant tissue. Nature, $215:363$.

17. Lyttleton, J. W. 1962. Isolation of ribosomes from spinach chloroplasts. Expl. Cell Res., $26:312$.

18. Richards, O. C. 1967. Hybridization of *Euglena gracilis* chloroplast and nuclear DNA. Proc. Natl. Acad Sci., $57:156$.

19. Scott, N. S., and R. Smillie. 1967. Evidence for the direction of chloroplast ribosomal RNA synthesis by chloroplast DNA. Biochem. Biophys. Res. Comm., 28:598.

20. Spencer, D., and P. R. Whitfield. 1967. Ribonucleic acid synthesizing activity of spinach chloroplasts and nuclei. Arch. Biochem. Biophys., 121:336.

21. Spencer, D., and P. R. Whitfield. 1967. DNA synthesis in isolated chloroplasts. Biochem. Biophys. Res. Comm., 28:538.

22. Spencer, D., and S. G. Wildman. 1962. Observations on the structure of grana-containing chloroplasts and a proposed model of chloroplast structure. Australian J. Biol. Sci., 15:599-610.

23. Stutz, E., and H. Noll. 1967. Characterization of cytoplasmic and chloroplast polysomes in plants: Evidence for three classes of ribosomal RNA in nature. Proc. Natl. Acad. Sci., 57:774.

24. Tewari, K. K., and S. G. Wildman. 1966. Chloroplast DNA from tobacco leaves. Science, 153:1260-1271.

25. Tewari, K. K., and S. G. Wildman. 1967. DNA-dependent RNA polymerase of tobacco chloroplasts. Fed. Proc., 26:869.

26. Tewari, K. K., and S. G. Wildman. 1967. DNA polymerase in isolated tobacco chloroplasts and the nature of the polymerized product. Proc. Natl. Acad. Sci., 58:689-696.

27. Tewari, K. K., and S. G. Wildman. 1968. Function of chloroplast DNA. I. Hybridization studies involving nuclear and chloroplast DNA with RNA from cytoplasmic (80S) and chloroplast (70S) ribosomes. Proc. Natl. Acad. Sci., 59:569-576.

28. Wildman, S. G. 1967. The organization of grana-containing chloroplasts in relation to location of some enzymatic systems concerned with photosynthesis, protein synthesis, and ribonucleic acid synthesis. In *Biochemistry of Chloroplasts*, T. W. Goodwin, ed., Vol. II, pp. 295-320. London: Academic Press.

29. Wildman, S. G., T. Hongladarom, and S. I. Honda. 1962. Chloroplasts and mitochondria in living plant cells: Cinephotomicrographic studies. Science, 138:434-436.

Functional Autonomy and Evolution of Organelles as Reflected in Ribosome Structure

HANS NOLL
Department of Biological Sciences
Northwestern University, Evanston, Illinois

Editor's note: The manuscript for this paper was not available. The following summary is adapted from the taped proceedings.

THE QUESTION OF ORGANELLES, and specifically mitochondrion organization and replication, was of long-standing interest to Dr. Noll. Having completed some major work on protein synthesis in rat liver, he was aroused by the discovery that mitochondria and chloroplasts contained DNA, and being brought up on the right dogmas—DNA makes RNA, and RNA directs protein synthesis—he found it rather attractive to look into a new type of protein synthesizing system. He was encouraged to do so by several circumstances. First of all, he had spent considerable time developing the machinery for running high resolution sucrose gradient work, and was eager to train it on objects that were hard to resolve by the old methods. Second, he was able to convince a very competent chloroplast biochemist, E. Stutz, to join his laboratory. The other circumstance was a challenge presented by some work on the subject by Brawerman and Eisenstadt (1964), who described the presence within chloroplasts of a 60 S ribosome, something unusually small never seen before, and of 19 S ribosomal RNA not only in the cytoplasm, but in the chloroplasts as well. These figures suggested they may be dealing with degradation products, and it became a challenge to show that this system must be in some way classical or orthodox.

The high resolution gradient analyzing system that was used has been described elsewhere (Noll, 1969). It is a constant flow system in which distilled water is pumped onto the top of the gradient in the tube. The bottom of the tube is punctured and the flow regulated so that it is slower than if the tube were drained by gravity. The rate for small gradients generally used was one half ml per minute. The size of the flow cell is critical and is designed for nonturbulent flow. Fractions can be collected in an ice-cooled automatic fraction collector that is time operated to be synchronous with the pumping action.

Structure and function of ribosomes

Before describing the experimental work, it is necessary to summarize some of the pertinent data on ribosome structure and function. The ribosome is a particle which has a bipartite design, consisting of a large and a small subunit. Both subunits contain RNA, and in the *E. coli* system, the large one sediments with 50 S and the smaller one with 30 S. The messenger is attached to the smaller subunit. At low Mg concentration these dissociate into separate subunits, which can be broken down further with the detergent sodium dodecylsulfate (SDS) and the RNA extracted. The larger subunit yields 23 S RNA and the smaller one 16 S RNA. The protein that is integrated with this RNA is a tightly knit structure which consists of about 30 to 40 different molecules of protein in the 50 S subunit and about 20 to 25 in the smaller. These proteins can be resolved by means of acrylamide gel electrophoresis into a highly characteristic system of bands that can be used as a fingerprint.

The function of the ribosome is to translate messages. The initiation of protein synthesis begins with the 30 S subunit, which first accepts a messenger. There are two binding sites, which we can designate as A and B. In *E. coli* the initiator triplet on the messenger is AUG, which codes for the n-formylated form of methionine, and which we will call met-F. The met-F tRNA is then inserted into the first site A to form the initiator complex. Several different factors are necessary, including a binding factor which must recognize three sites: the AUG triplet, the met-F tRNA, and the A site of the 30 S subunit. When the initiation complex is formed and the 50 S subunit is added, the next step is to move the messenger over to the next site B, so that the second triplet can be inserted into the decoding site A. This act requires the hydrolysis of GTP, the energy of which is used to move the messenger over by one triplet.

If the second triplet is, for example, GUU which codes for valine, then the site will now attract valine-tRNA. The next step is to form a peptide bond between the met-F and valine, and then hydrolyze the ester bond that holds the met-F to the first aminoacyl-tRNA. The first peptide bond is now completed, and from then on the process continues in the same way; as soon as the peptide bond has been formed, the messenger is translocated by one triplet and the next amino acid is added.

One very important conclusion that can be drawn from all this is that there is a complementarity between the initiating triplet, the 30 S subunit, and the messenger binding factor, thus providing an interesting example of ribosomal specificity. This is important for the question of what messages can be translated by a given ribosome, for apparently not all messages can be translated by any given ribosome. Thus, if a mammalian ribosome is presented with a bacterial messenger, it will not be able to translate it because it does not respond to AUG, or so we believe. The stability of the ribosome, messenger, and nascent protein complex depends, according to Noll's model, on the length of the protein chain and its degree of coiling once it leaves the confines of a groove through the ribosome.

Comparative studies of cytoplasmic and organelle ribosomes

The main objective of the experimental work was to determine the relative sizes of ribosomes obtained from different sources. Work of Lyttleton (1962) indicated that chloroplasts had 66 S ribosomes. Presumably *E. coli* had 70 S ribosomes, while rat liver had 80 S ribosomes, yet the differences were far from clear, since nobody had ever run the different types on a sucrose gradient under the same conditions. Thus the test for difference adopted here was the ability to resolve the various types after mixing together reasonably pure preparations of each.

The first investigations immediately showed that chloroplast ribosomes were indeed different from the cytoplasmic ribosomes (Stutz and Noll, 1967). Using bean ribosomal preparations from cytoplasm and from chloroplasts showed a 70 S peak for the chloroplast ribosomes and an 80 S peak for the cytoplasmic. When the two were mixed together, the 70 S and 80 S peaks were clearly resolved.

The studies were extended to compare plant and animal material. Rat liver cytoplasmic and bean cytoplasmic ribosomes run together on a sucrose gradient could not be resolved, indicating almost identical sedimentation constants. Similarly, a mixture of *E. coli* ribosomes and bean chloroplast ribosomes again could not be resolved, indicating very similar sedimentation constants. Finally, a mixture of bean cytoplasmic and *E. coli* ribosomes were clearly resolved. However, the separation was not quite as great in this case as when chloroplast ribosomes were run against cytoplasmic, indicating that chloroplast ribosomes are slightly smaller than those of *E. coli,* agreeing with the previously determined value of 68 S.

In the next experiments the results were somewhat unexpected. Although it had long been assumed that the ribosomes from the cytoplasm of plants and animals were the same, there were nevertheless differences which appeared if one looked into the ribosomal RNA of the subunits. If we compare rat liver, bean cytoplasm, *E. coli,* and chloroplasts, we find that both rat liver and bean have an 80 S ribosome, *E. coli* has 70 S, and chloroplasts 68 S. However, if rat liver and bean preparations are extracted with SDS, the resulting ribosomal RNA from the rat liver sediments with 29 S and 18 S, which is clearly different from the bean 25 S and 17 S. The *E. coli* with 23 S and 16 S and the chloroplast with 23 S and 16 S are indistinguishable. This indicated that plants really had different ribosomes from those of animals, a fact that was not known before.

The situation in *Euglena* again showed 70 S chloroplast ribosomes with subunits of 50 S and 30 S (Rawson and Stutz, 1968). The cytoplasmic ribosomes, however, showed a sedimentation constant of 86 S, somewhat larger than rat, and clearly resolvable if rat liver and *Euglena* cytoplasmic ribosomes were mixed. The ribosomal RNA is also somewhat unorthodox. While the chloroplast ribosomes sediment with 22 S and 16 S, very similar to other chloroplast ribosomes in nature, the pattern is somewhat different for the cytoplasmic ribosomes. Here the 24 S and 20 S RNA is unusually large, the largest ribosomal RNA in any subunit from any organism so far.

Work on mitochondrial ribosomes was carried out on preparations from *Neurospora* (Küntzel and Noll, 1967). Here the separation was not easy, for the cytoplasmic ribosomal peak is about 77 S and the mitochondrial is 73 S, and again the test was that relatively pure preparations of each could be resolved into two peaks when mixed together. The subunit RNA also showed differences. The cytoplasmic rRNA, which is about 25 S and 17 S is characteristic of plants. The mitochondrial rRNA is about 23 S and 16 S, as found for bacteria. A study of the base composition of the mitochondrial rRNA also clearly showed it to be different from the cytoplasmic. Work reported from the laboratory of David Luck, however, has not been able to distinguish between the 77 S cytoplasmic and the 73 S mitochondrial ribosomes, and this discrepancy has not as yet been settled.

A scheme for ribosomal evolution

Summarizing the data presented so far, we find two main groups: the 80 S group, including rat liver, and the slightly smaller bean and *Neurospora* cytoplasmic ribosomes; and the 70 S group extending from 68 S to 73 S, including *E. coli*, chloroplasts, and mitochondria. These results, along with data from other sources, can now be arranged in a highly speculative scheme of ribosomal evolution. Starting from some unknown precursor, there is a group of procaryotes, consisting of bacteria, chloroplasts, and mitochondria, all characterized by ribosomal RNA which is 23 S - 16 S. Branching from this main stem are the higher plants, which are 25 S - 17 S. The same figure also applies to some primitive eucaryotes like *Paramecium* and *Tetrahymena*. Then there are the anomalies, *Euglena* with 24 S - 20 S and *Amoeba* with 26 S - 20 S. One could suggest that this class represents a group of evolutionary deviants, dead-end products of evolutionary experimentation. On the other hand, the insects still belong to the 25 S - 17 S class which we find in higher plants. However, ascending the evolutionary tree, the size of the ribosomes increases quite clearly. The birds have 27 S - 18 S, and the mammals are at the top with 29 S - 18 S. The increase in size is especially noticeable in the larger subunit, which increases from 23 S to 29 S. There is a lesser, but also distinct increase in the small subunit from 16 S to 18 S.

One might question the significance of these size determinations and ask whether one could not find relationships, for example, in the composition of the proteins or in the RNA base sequences. However, the great variability in the proteins, even from one species to another, as well as variations in RNA base sequences, reveal no clear pattern, and size therefore seems to be one of the best guides we have at present.

Autonomy of mitochondria

If we now consider the question of the origin of mitochondria, the difficulty is that the DNA in mitochondria is rather small. Therefore, the idea that mitochondria are autonomous has to contend with the fact that a large portion of the mitochondrion structure must be coded for by nuclear DNA. What is the evidence at present? We know for certain only that a structural protein of the inner membrane of the mitochondrion is synthesized by mitochondrial ribosomes. On the other hand, as Dr. Nass reported earlier, there is evidence from hybridization studies that the ribosomal RNA, as well as at least some of the tRNA, are coded for by the mitochondrial DNA. If we sort out those objects we know are made by the mitochondrial DNA, this would include all of the ribosomal RNA. Since the combined ribosomal RNA weight is one and a half million, this would already account for three million of the DNA weight. If the DNA weight is on the order of ten million, we have already used up one third. If we try to code for the tRNA we might use up about another third, and the remaining third is clearly not enough to code for the ribosomal proteins. If the ribosomal proteins are made up of about fifty different molecules and we take a minimum molecular weight of 12,000, about ten million molecular weight units of DNA are needed for that alone.

We are now faced with the question of where the mitochondrial ribosomes are made. Since we know that the rRNA is transcribed on mitochondrial DNA, we are then forced to the conclusion that the proteins that make up the ribosomes must be synthesized on nuclear messenger. Here we have one tool that can be used to solve this problem, namely, the differential sensitivity of mitochondrial ribosomes to antibiotics. The various classes of ribosomes that exist in nature are distinct in their response to antibiotics. There is a large class of antibiotics that inhibits only bacterial type ribosomes, including mitochondrial and chloroplast ribosomes, the best known of which are streptomycin and chloramphenicol. Antibiotics which are specific for the 80 S type ribosomes are tetracyclins and cyclohexamide.

Let us consider the two alternatives, which we will call A and B. In pathway A, nuclear messenger that codes for mitochondrial ribosomal proteins forms polysomes in the cytoplasm in the classical fashion with cytoplasmic ribosomes. As the ribosomal proteins are made, they must penetrate the mitochondrial membranes and find their place on the nascent or finished ribosomal RNA that has been transcribed on mitochondrial DNA. They can now aggregate, fold up, and then the newly formed mitochondrial ribosome is able to engage in the process of translation on a messenger made on the mitochondrial DNA.

The second alternative, B, in which mitochondrial ribosomes are programmed with nuclear messenger, requires that messenger from the nucleus is transported into the mitochondrion, so that nuclear messenger, instead of being translated in the cytoplasm on cytoplasmic ribosomes, would be translated in the mitochondrion on mitochondrial ribosomes. The nuclear messenger might be transported into the mitochondrion by some sort of direct contact between mitochondrion and nucleus, or it might possibly be wrapped in a protein coat to protect it and travel through the cytoplasm to the mitochondrion. These particles would then infect the mitochondrion at a particular site on its surface, very much like an RNA virus infects a bacterium.

How could we distinguish between the two alternatives? If A is correct, antibiotics which specifically inhibit cytoplasmic translation would inhibit the labeling or the synthesis of mitochondrial ribosomes. Likewise, antibiotics like chloramphenicol that specifically inhibit mitochondrial ribosome translation would not inhibit the synthesis of mitochondrial ribosomal proteins. If scheme B is correct, there should be no inhibition of the synthesis of mitochondrial ribosomal proteins with cytoplasmic inhibitors like cyclohexamide, but there would be inhibition with the bacterial type of inhibitors like chloramphenicol. There is one difficulty, however, for if the messenger is wrapped up in a protein particle with a fast turnover, it is conceivable that the cytoplasmic inhibitor cyclohexamide would still inhibit the synthesis of bacterial type of mitochondrial ribosomes. Therefore this test would not distinguish between these two situations. On the other hand, chloramphenicol should still prevent the synthesis of mitochondrial ribosomes, because even though the messenger can get in, it cannot be translated.

Recently Küntzel (1969) has reported some experiments using these antibiotics, which favor scheme A, that is, that ribosomal proteins of mitochondria are synthesized in the cytoplasm, not only on nuclear messenger, but also on cytoplasmic ribosomes. On the other hand, evidence in direct conflict with this has been reported from Luck's laboratory, presumably showing that mitochondrial ribosomal proteins are coded for by mitochondrial DNA. Obviously the conflict is not yet resolved. Of the two pathways, B would seem more plausible since there are many precedents for this sort of transport of messenger from one place to another. It is a mechanism that normally takes place in cytoplasmic protein synthesis and also infection by RNA phages, but there is no precedent for the type of situation in scheme A. It is possible also that the size of mitochondrial DNA in *Neurospora,* which was used in these experiments, is a good deal larger than that in mammalian mitochondria. Whatever the case, a great deal more work and some new ideas are needed to resolve these issues.

Literature Cited

Brawerman, G., and J. M. Eisenstadt. 1964. Template and ribosomal ribonucleic acids associated with the chloroplasts and the cytoplasm of *Euglena gracilis*. J. Mol. Biol., *10:*403-411.

Küntzel, H. 1969. Proteins of mitochondrial and cytoplasmic ribosomes from *Neurospora crassa.* Nature, *222:*142-146.

Küntzel, H., and H. Noll. 1967. Mitochondrial and cytoplasmic polysomes from *Neurospora crassa.* Nature, *215:*1340-1345.

Lyttleton, J. W. 1962. Isolation of ribosomes from spinach chloroplasts. Exptl. Cell Res., *26:*312-317.

Noll, H. 1969. An automatic high-resolution gradient analyzing system. Analyt. Biochem., *27:*130-149.

Rawson, J. R., and E. Stutz. 1968. Characterization of *Euglena* cytoplasmic ribosomes and ribosomal RNA by zone velocity sedimentation in sucrose gradients. J. Mol. Biol., *33:*309-314.

Stutz, E., and H. Noll. 1967. Characterization of cytoplasmic and chloroplast polysomes in plants: evidence for three classes of ribosomal RNA in nature. Proc. Nat. Acad. Sci., *57:*774-781.

Appendix

Thirtieth Annual Biology Colloquium

Theme: **Biological Ultrastructure:** The Origin of Cell Organelles

Date: April 25-26, 1969

Place: Oregon State University, Corvallis, Oregon

Standing Committee for the Biology Colloquia: Paul O. Ritcher, *chairman,* Malcolm E. Corden, Ernst J. Dornfeld, Paul R. Elliker, Gwil O. Evans, Henry P. Hansen, Hugh F. Jeffrey, J. Kenneth Munford, and John M. Ward

Special Committee for the 1969 Biology Colloquium: Thomas C. Allen, *co-chairman,* Patricia J. Harris, *co-chairman,* Robert F. Dyson, Dorothy K. Fraser, John E. Morris, and Ralph S. Quatrano

Colloquium Speakers, 1969:

John H. Luft, professor, Department of Biological Structure, University of Washington Medical School, Seattle, *leader*

Roderic B. Park, professor, Department of Botany, University of California, Berkeley

Margit M. K. Nass, associate professor, Department of Therapeutic Research, University of Pennsylvania School of Medicine, Philadelphia

Germaine Cohen-Bazire, associate research bacteriologist, Department of Bacteriology and Immunology, University of California, Berkeley

Samuel G. Wildman, professor, Department of Botanical Sciences and Molecular Biology Institute, University of California, Los Angeles

Hans Noll, professor, Department of Biological Sciences, Northwestern University, Evanston

Cooperating Organizations:

National Science Foundation
School of Science, Oregon State University
School of Agriculture, Oregon State University
Phi Kappa Phi
Sigma Xi
Omicron Nu
Phi Sigma